AHS Rotorcraft Handling Qualities Technical Meeting 2017

Huntsville, Alabama, USA
22 - 23 February 2017

ISBN: 978-1-5108-4676-0

Printed from e-media with permission by:

Curran Associates, Inc.
57 Morehouse Lane
Red Hook, NY 12571

Some format issues inherent in the e-media version may also appear in this print version.

Printed by Curran Associates, Inc. (2017)

American Helicopter Society International
2701 Prosperity Ave, Suite 210
Fairfax, VA 22031

Phone (703) 684-6777
Fax: (703) 739-9279

staff@vtol.org

Additional copies of this publication are available from:

Curran Associates, Inc.
57 Morehouse Lane
Red Hook, NY 12571 USA
Phone: 845-758-0400
Fax: 845-758-2633
Email: curran@proceedings.com
Web: www.proceedings.com

TABLE OF CONTENTS

Investigation of a Compound Helicopter Flying the Depart and Abort Mission Task Element

Kevin Ferguson
Kevin.Ferguson@selex-es.com
Research Assistant
University of Glasgow
Glasgow, United Kingdom

Douglas Thomson
Douglas.Thomson@glasgow.ac.uk
Senior Lecturer
University of Glasgow
Glasgow, United Kingdom

Dave Anderson
Dave.Anderson@glasgow.ac.uk
Senior Lecturer
University of Glasgow
Glasgow, United Kingdom

ABSTRACT

The next generation of rotorcraft will have to satisfy the appropriate handling qualities requirements before entering service. Many of these vehicles will operate at significantly greater speeds than the conventional helicopter and will therefore have different capabilities than current helicopters. Due to the different capabilities of the compound helicopter, it is possible that new Mission Task Elements (MTEs) need to be developed to assess the handling qualities of this type of helicopter. It is also possible that existing MTEs may be suitable without modification. Overall, it seems necessary to review the US Army's current handling qualities specification, ADS-33, and determine the suitability of the current MTEs for compound vehicles. The broad aim of the paper is to assess the performance of compound helicopter during manoeuvring flight. More specifically, a simulation study of a compound helicopter flying the Depart and Abort ADS-33 Mission Task Element. There are two objectives: firstly the capabilities of the compound vehicle is compared with those of a conventional helicopter, and secondly, the suitability of the current Depart and Abort MTE, for compound vehicles, is assessed. The results of the research study highlight the capability of compound helicopters in low speed acceleration manoeuvres. These results can be used to redefine low speed acceleration manoeuvres in the new update to the ADS-33 specification. The results also indicate some information about the potential design issues with the compound helicopter.

NOMENCLATURE

Main Nomenclature List

α_{bl}	blade angle of attack (rad)
χ	track angle (rad)
δ_c	collective lever or lateral stick position (%)
δ_s	longitudinal stick position (%)
δ_{tr}	foot pedal position (%)
ε	blade angle of attack where the maximum lift-to-drag ratio occurs
γ	glideslope angle (rad)

Presented at the AHS Rotorcraft Handling Qualities, Huntsville, Alabama, February 22–23, 2017.

γ_{rot}	main rotor Lock number
ρ	air density (kg/m^3)
Γ	tail-rotor cant angle (deg)
σ	main rotor solidity
θ_0	main rotor collective angle (rad)
θ_{0tr}	tail-rotor collective angle (rad)
θ_{1s}	main rotor longitudinal cyclic angle (rad)
θ_{1c}	main rotor lateral cyclic angle (rad)
θ_{prop}	Cheyenne propeller pitch angle (rad)
θ_{tw}	blade twist (rad)
e	Oswald efficiency factor
f_e	equivalent flat plate area (m^2)
m	aircraft mass (kg)
$\mathbf{u}$	control vector (rad)
$\mathbf{y}_{des}$	desired aircraft trajectory (various units)
x_e, y_e, z_e	aircraft position in Earth axes (m)
AR	aspect ratio
Ω	main rotor rotational speed (rad/s)
Φ, Θ, Ψ	Euler angles (rad)
C_D, C_L	drag and lift coefficient
C_T	thrust coefficient
D	drag (N)
L_w	wing lift (N)
$I_{xx}, I_{yy}, I_{zz}, I_{xz}$	moments of inertia (kg.m^2)
P	power (kW)
R	main rotor radius (m)
S	wing area (m^2)
T	thrust (N)
V_f	airspeed (kt or m/s)
W	aircraft weight (N)
X	propeller thrust at trim (N)

Subscripts

prof	profile drag
prod	produced
prop	propeller
ind	induced drag
req	required
fin	fin
flat	flat plate
fus	fuselage
rot	rotor
tr	tail-rotor
tw	twist
w	wing

INTRODUCTION

The compound helicopter is again being explored as it can potentially satisfy the emerging requirements for the next generation of rotorcraft. The US Army started their Joint Multi-Role (JMR) programme in 2014, which aims to replace their existing fleet with helicopters with greater speed and range capability (Ref. 1). As a result, some helicopter manufacturers are investigating the compound helicopter design, as it can potentially satisfy the requirements of the JMR programme. For example, Sikorsky have recently concluded their testing of the Sikorsky X2, with the flight tests indicating promising results. The testing of the Sikorsky X2 programme was so successful that Sikorsky have taken this concept to the next level by developing the Sikorsky S-97 Raider®. The Sikorsky S-97 Raider is envisioned to be a multi-role aircraft which can complete various military missions including close-air support and armed reconnaissance. In addition, Airbus Helicopters have flight tested their prototype - the Airbus Helicopters X^3. It is clear that some helicopter manufacturers are investigating the compound helicopter design, as it can potentially satisfy the requirements of the JMR programme. It is also clear that compound helicopter designs will operate at significantly greater speeds than the conventional helicopter and will therefore have different capabilities than current helicopters. Due to the different capabilities of the compound helicopter, it is possible that new Mission Task Elements (MTEs) need to be developed to assess the handling qualities of this type of helicopter. It is also possible that existing MTEs may be suitable without modification. Overall, it seems necessary to review the US Army's current handling qualities specification, ADS-33, and determine the suitability of the current MTEs for compound vehicles.

The broad aim of the collaboration between the University of Glasgow and the US Army's Research, Development and Engineering Command (RDECOM) is to assess the performance of compound helicopters during manoeuvring flight. More precisely, this study focuses on the Depart and Abort MTE and examines how a compound helicopter would fly this particular test manoeuvre. There are two main objectives from the study. The first objective is assess the capabilities of the compound vehicle and compare these capabilities with a conventional helicopter. The second objective is to determine the suitability of the current Depart and Abort MTE for compound vehicles.

This research aim is achieved by using the flight dynamics tool of inverse simulation. Inverse simulation reverses the process of conventional simulation. With conventional simulation, a control input is applied to the mathematical model and a response is observed. Inverse simulation does the opposite and the manoeuvre of interest becomes the input to the simulation, and the vehicle response and pilot control activity required to fly it are computed. One of the benefits of inverse simulation is that it does not require detailed knowledge of the underlying model and is thus extremely effective for investigating performance and control strategies. To use the inverse simulation algorithm successfully requires three key elements: a mathematical model, a manoeuvre model and the inverse simulation algorithm itself. Each of these elements are discussed in the subsequent sections.

MATHEMATICAL MODELLING

For the majority of the study, the mathematical model used is a 6 degrees of freedom nonlinear model with a disc representation of the main rotor. This is a well established formulation (Ref. 2), which provides valid and usable results over a wide range of flight conditions. This model is referred to as the "level 1" model, which is consistent with the modelling hierarchy as defined by Padfield (Ref. 2). Comprehensive descriptions of this model can be found in the literature (Refs. 3–5), therefore only a short description is given here. The level 1 main rotor model ignores the pitching and lagging degrees of freedom, therefore assuming that the flap dynamics have the most influence in terms of the helicopter's flight dynamic characteristics. The flapping dynamics are assumed to be quasi-steady, a common assumption in main rotor modelling, therefore permitting a multi-blade representation of the main rotor. The main rotor is assumed to be centrally hinged with stiffness in flap and with the main rotor chord assumed to be constant. Furthermore, the model also features a steady momentum inflow and a rotor-speed governor model. One of the short comings of this type of model is that it is unable to model nonlinear aerodynamics. This limitation becomes important in high speed flight as compressibility effects and blade stalling need to be modelled to successfully capture the behaviour of the helicopter's main rotor (Ref. 6). To accommodate this modelling deficiency a new helicopter main rotor model was developed, which is referred to as the "level 2" model. The main difference between the two models is that nonlinear aerodynamics are modelled in the level 2 model. As a result, the level 2 model is able to model compressibility effects, reverse flow and blade stalling. It should be stressed that this is not a modelling study, therefore only a limited amount of time was dedicated to developing this level 2 model.

Fig. 1. The UH-60 "Blackhawk" Helicopter

Consequently, the majority of the results presented in this study were generated using the level 1 model. The validity of these types of models are discussed further in the report.

UH-60A Helicopter Model

The modelling strategy used in this study is to use an established mathematical model of a conventional helicopter, and then convert this model to represent a compound helicopter configuration. One of the first tasks in the study is to configure the mathematical model to represent a conventional helicopter. The conventional helicopter selected is the UH-60A, as shown in Figure 1. The UH-60A helicopter is a utility transport helicopter developed by Sikorsky, which performs various missions such as aeromedical evacuations and troop assaults. It is a twin-engined medium sized helicopter, designed for a gross weight of 7200 kg, and its main design parameters are shown in Table 1. The main rotor of the vehicle features four articulated rotor blades coupled with hydraulic lag dampers. The helicopter also features a canted tail-rotor which provides additional lift in the hover and extends the centre of gravity range of the aircraft (Ref. 7). Another key design feature is the moveable tailplane, with the incidence of the tailplane being a function of airspeed, collective lever position, pitch rate and lateral acceleration. The UH-60A configuration is used for a starting point in this study for two primary reasons. Firstly, the data-set is in the public domain (Refs. 8, 9) thereby allowing dissemination of results. Secondly, the UH-60A helicopter falls into the utility helicopter category, which is the type of helicopter the Joint Multi Role (JMR) programme will be developing.

Compound Helicopter Configuration

The next step is to take the conventional helicopter, Figure 1, and convert it to model a compound configuration. The compound configuration examined is a similar configuration to the Lockheed Cheyenne helicopter. This configuration features a wing and a pusher propeller mounted at the tail of the airframe. A sketch of this compound helicopter configuration is shown in Figure 2(b). The end result is two helicopter configurations: a conventional helicopter and a compound helicopter configuration.

Design Considerations

Wing Design The compound helicopter configuration have been introduced. The addition of wings and a pusher propeller to the helicopter, see Figure 2(b), means that some preliminary design issues need to be taken into account. The first task is to size the wing. Orchard and Newman investigated the design of compound helicopter (Ref. 10). Their study suggests that a medium size wing should be used to provide a compromise between the beneficial effect of offloading the rotor at high speeds and the adverse effect of creating an aerodynamic download at low speeds (Ref. 10). Their study also highlights the importance of the wing design to the vehicle, suggesting that an

Design Parameter	Value
m	7200 kg
R	8.18 m
Ω	27 rad/s
σ	0.08210
γ_{rot}	8.19
θ_{tw}	-18 deg (nonlinear)
I_{xx}	6290 kg.m^2
I_{yy}	51991 kg.m^2
I_{zz}	49895 kg.m^2
I_{xz}	2541 kg.m^2
R_{tr}	1.68 m
Ω_{tr}	125 rad/s
σ_{tr}	0.1875
$\theta_{tr\ tw}$	-18 deg
Γ	20 deg

Table 1. Important Design Parameters of the UH-60A Helicopter (Ref. 8)

aspect ratio of 6 would be appropriate so that the wing does not extend into a region where the the main rotor's tip vortex would adversely affect the wing's performance (Ref. 10). All of these issues should be taken into account when sizing the wing. Another important study by Yeo and Johnson (Ref. 6) concluded that the optimum cruise condition occurs when the wing carries 91-92% of the aircraft weight. It also makes sense to design the wing so that it operates at its maximum lift-to-drag ratio at cruise. The assumed cruise condition of the compound helicopter is 210kt at an altitude of 1800m. These two values have been loosely based on the requirements of the JMR programme (Ref. 1). The estimated wing area is found by using the following Equation

$$S = \frac{2L_w}{\rho V^2 C_L} \tag{1}$$

where L_w is the lifting force of the wing (assumed to be 91% of the aircraft's weight), and the other terms retain their usual meaning. The lift coefficient corresponds to the condition where the wing's aerofoil lift-to-drag ratio is at its highest. For the compound helicopter the NACA-65A-618 aerofoil is used due to its low profile drag at trim, its relatively high lift coefficient, gentle stall characteristics and its high lift curve slope (Ref. 11). For this particular aerofoil the maximum lift-to-drag ratio occurs when $C_L = 0.6$. Using this lift coefficient and the appropriate cruise flight conditions then the wing area, S of the compound helicopter is calculated to be 17.6 m^2. This choice of wing area is a design compromise between minimising the hover penalty of the wing whilst having the ability to offload the main rotor in high speed flight.

Another design decision is the selection of the aspect ratio. The choice of aspect ratio was based on historical data of winged helicopters which have previously flown, which include the NH-3A and SA341 Gazelle helicopters (Refs. 12, 13), leading to the design choice of 6. This choice is also supported by Orchard and Newman (Ref. 10), as previously stated. The taper ratio of each wing is selected to be 0.5, in an attempt to reduce the amount of induced drag (Ref. 11).

Main Rotational Speed Variation Due to the expansion of the flight envelope, it is necessary to reduce the rotational speed of the compound helicopter's main rotor in high speed flight (Ref. 10). This is required to avoid adverse compressibility effects in forward flight. This design feature is incorporated on all modern compound helicopters. In this study, the main rotor planform of each helicopter configuration is the same. The only difference between the main rotor design of the conventional helicopter when compared to the compound helicopter is the rotational

(a) Baseline Helicopter

(b) Cheyenne Compound Helicopter

Fig. 2. Sketches of the two Helicopter Configurations

speed in forward flight. Figure 3 shows how the rotational speed of the compound helicopter's main rotor varies with airspeed. Note that the rotorspeed of the UH-60A helicopter is a constant value of 27 rad/s throughout the flight envelope. In relation to the compound helicopter, at 130 kt the main rotor's speed is gradually lowered to avoid the assumed drag divergence Mach number. The drag divergence Mach number is the Mach number where the drag coefficient rises significantly (Ref. 7). This is commonly defined as the Mach number were the drag coefficient rises at a rate of 0.1 per unit Mach number (Ref. 7). Experimental results from various studies confirm that the drag divergence Mach number for typical helicopter aerofoils ranges between 0.85-0.92 (Refs. 14, 15). The value of 0.87 is selected for this study. At 210 kt, the main rotor is slowed to 81% (see Figure 3), thereby avoiding an advancing tip Mach number of 0.87. The design decision here is to keep the rotorspeed as high as possible, without compressibility effects becoming limiting, to avoid vibration issues due to the lowering of the blade passing frequency (Ref. 10).

Propeller Design To attempt to optimise the design of the propeller, for some given flight condition, a simple design algorithm was developed. This design algorithm is largely based on the work of Adkins and Liebeck (Ref. 16). The aim of the algorithm is to output a propeller design with the greatest achievable propulsive efficiency at some flight condition. Like the algorithm of Adkins and Liebeck (Ref. 16), the user is required to input the radius of the propeller, its rotational speed and the amount of thrust the propeller is required to generate. Also, the user is required

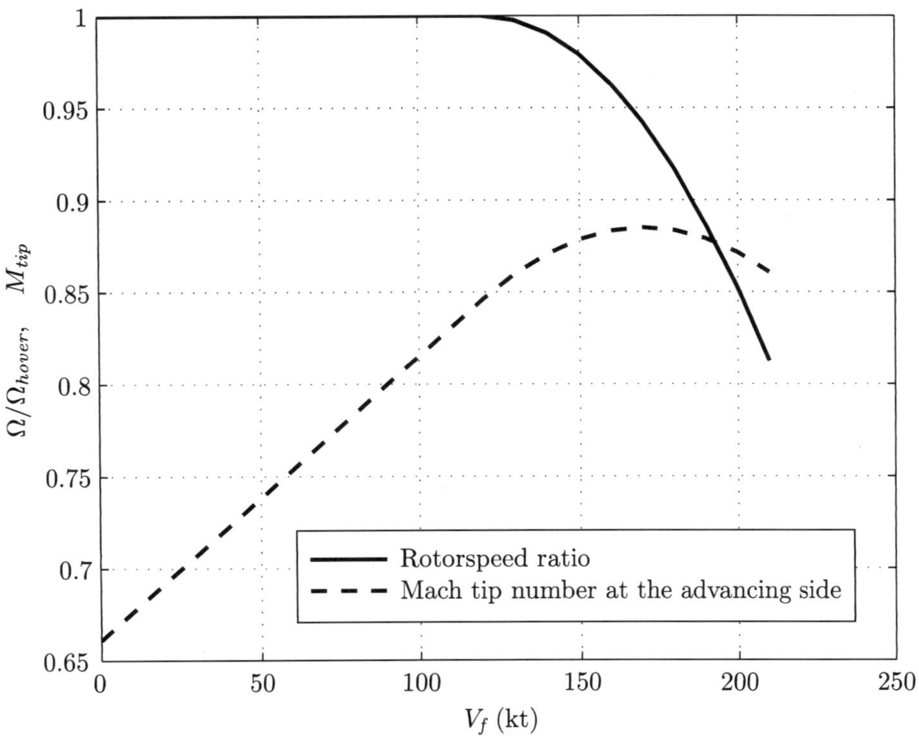

Fig. 3. Variation of Rotorspeed for the compound helicopter

to input the appropriate flight conditions such as airspeed, altitude density and glideslope. Essentially, the algorithm calculates two design parameters which are of critical importance to the propeller. The first parameter is the mean chord (or solidity) of the propeller to produce the required thrust. It is well understood that the profile power of the propeller is kept to a minimum when each blade section operates at an angle of attack where the local lift-to-drag ratio is maximised (Refs. 16, 17). Therefore, the second parameter calculated is the propeller twist to minimise profile power. The functions that the algorithm solves are:

$$T_{req} - T_{prod} = 0 \tag{2a}$$

$$\varepsilon_1 - \alpha_1 = 0 \tag{2b}$$

$$\varepsilon_2 - \alpha_2 = 0 \tag{2c}$$

$$\vdots$$

$$\varepsilon_n - \alpha_n = 0 \tag{2d}$$

where T_{req} is the required propulsive thrust and T_{prod} is the thrust produced by the blade element propeller model. The term ε is the local angle of attack where the maximum lift-to-drag ratio is achieved. The local angle of attack at each propeller blade station is given by α. The subscripts in the ε and α terms represents the radial station of interest. For example, α_1 is the angle of attack at the propeller station nearest the propeller hub. Note that n corresponds to the number of blade elements considered. Hence, α_n is the angle of attack at the section nearest the propeller tip.

Cheyenne Compound Propeller At the cruise condition the propeller of the Cheyenne Compound is required provide all of the propulsive thrust, as the main rotor is significantly offloaded. For a conventional helicopter the drag of the airframe may be approximated with

$$D_{af} = D_{fus} + D_{rot} + D_{tr} + D_{fin} + D_{tp} \tag{3}$$

Calculating the drag of each component can be a difficult challenge, therefore numerical predictions of helicopter drag are usually carried out using the equivalent flat plate area. Using this method the drag of the airframe, D_{af}, can be stated as

$$D_{flat} = \frac{1}{2}\rho V_f^2 f_e \tag{4}$$

where f_e is the equivalent flat plate area of the helicopter. Equation (4) can be used to predict the drag of the compound helicopter, however the addition of other aircraft components needs to be taken into account. Hence, the drag of the Cheyenne compound can be expressed as

$$D_{tot} = D_{flat} + D_{w,\ prof} + D_{w,\ ind} \tag{5}$$

where $D_{w,\ prof}$ and $D_{w,\ ind}$ represent the profile and induced drag of the wing, respectively. The assumption made here is that the compound helicopter will feature a modern airframe design which reduces the amount of drag. Ormiston suggests (Ref. 18) that the equivalent flat plate area of modern designs can be approximated with

$$D_{flat} = 0.087\left(\frac{W}{1000}\right)^{2/3} \tag{6}$$

Using the aircraft mass of 7200 kg, then the flat plate area of the conventional helicopter is 1.49 m^2. The profile drag of the wing can be determined using a standard drag estimation technique, which is described in various design textbooks (Refs. 19, 20). The profile drag of the wing is given by

$$D_{w,\ prof} = \frac{1}{2}\rho V_f^2 S C_{D0} \tag{7}$$

where the amount of drag is captured by the zero lift drag coefficient, C_{D0}. Using the standard drag estimation method, the drag coefficient is approximated to be 0.01. The induced drag of the wing, $D_{w,\ ind}$, can be estimated using

$$D_{w,\ ind} = \frac{1}{2}\rho V_f^2 S C_{Di} \tag{8}$$

where C_{Di} is the induced drag coefficient which can be approximated using

$$C_{Di} = \frac{1}{AR\pi e} C_L^2 \tag{9}$$

The term e is the "Oswald efficiency factor", which is assumed to be 0.75. Using the cruise flight conditions then the total estimated airframe drag of the Cheyenne compound is 12600 N. Hence, at the cruise condition the pusher propeller is required to provide 12600 N of propulsive thrust. Typically, the highest achievable propulsive efficiency is about 0.86. Consequently, the goal here is to design a propeller with a propulsive efficiency, $\eta = TV/P$, of 0.86 at the cruise condition. Therefore, for the Cheyenne compound design the desired propeller power at cruise is 1582 kW.

Before applying the propeller design algorithm, the propeller's radius and rotational speed need to be determined. The propeller radius is determined by using historical data of modern propellers (Ref. 21). Marinus and

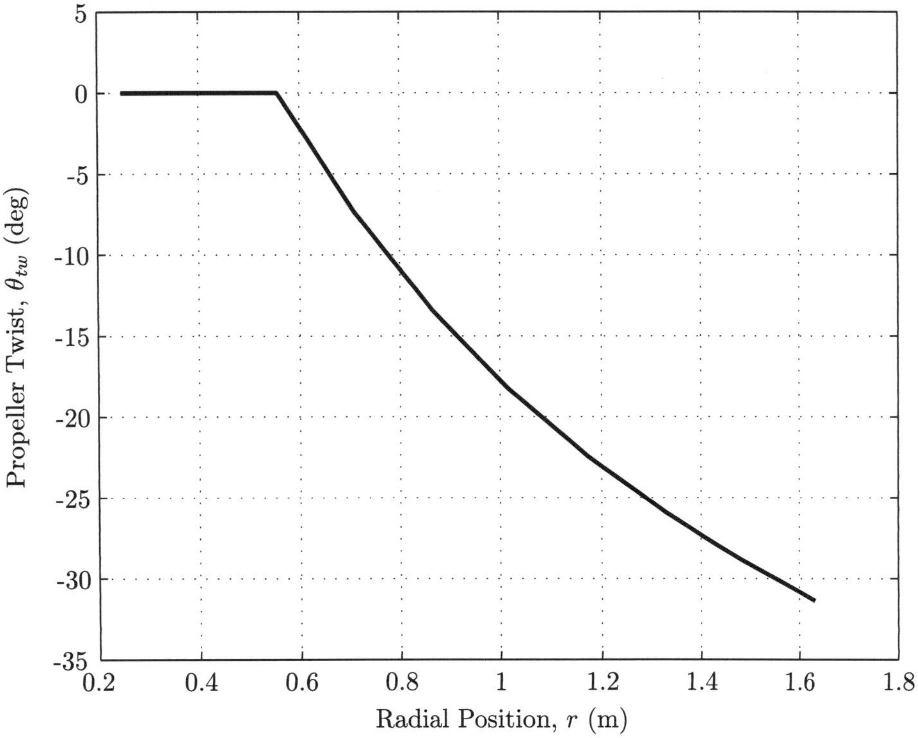

Fig. 4. Calculated Propeller Twist Distribution for the Cheyenne Compound

Poppe present the historical relationship between the diameters of different propellers and the maximum shaft power (Ref. 21). The maximum shaft power for the Cheyenne configuration is assumed to be 20% greater than the power required at cruise, to accommodate for emergency situations and aggressive manoeuvring flight. This gives a maximum shaft power of 1898 kW, therefore the propeller radius is predicted to be 1.71 m. The Mach tip number of the propeller should not exceed 0.87 at cruise, therefore the propeller's rotational speed is chosen to be 150 rad/s. Another design consideration is the choice of aerofoil sections. For this propeller design, the aerofoil selected is the NACA0012 aerofoil. The motivation for this aerofoil choice is that comprehensive "look-up tables" are available which allow the lift and drag coefficients to be calculated based on the local angle of attack and Mach numbers.

Some of the basic design parameters of the Cheyenne compound configuration have now been determined. Figure 4 shows the calculated optimum propeller twist distribution from the propeller design code. As expected, the propeller requires a large amount of negative twist to ensure a high propulsive efficiency at the cruise condition. The tangential velocity of each radial station changes significantly across the propeller blade, which requires a large amount of twist to optimise the angle of attack at each section. The tip of the propeller blade is pitched down by an angle of 31 deg, Figure 4. The variation of twist is nonlinear, although it could be approximated to be linear if desired. Note it is assumed that the inner portion of the propeller blade, between the root and 32% across the blade, has zero twist. The thrust capability of this part of the blade is limited, therefore no attempt was made to optimise the twist at this portion of the blade.

Figure 5 shows the resultant angle of attack variation across the propeller blade at the cruise condition. The blade angle of attack, α_{bl}, changes in a linear manner towards the tip, Figure 5. Note that towards the root there is a stall region due to the low local dynamic pressure, therefore these angles of attack are not shown in Figure 5. The maximum lift-to-drag ratio of each radial station is dependent on the local Mach number, due to the effects of compressibility. It is well known that the drag coefficient of a typical aerofoil increases dramatically as the local Mach number approaches unity (Ref. 7). As a result, the lower the Mach number, the higher the angle of attack where the maximum lift-to-drag ratio is achieved. This is shown in Figure 5, with the highest angle of attacks

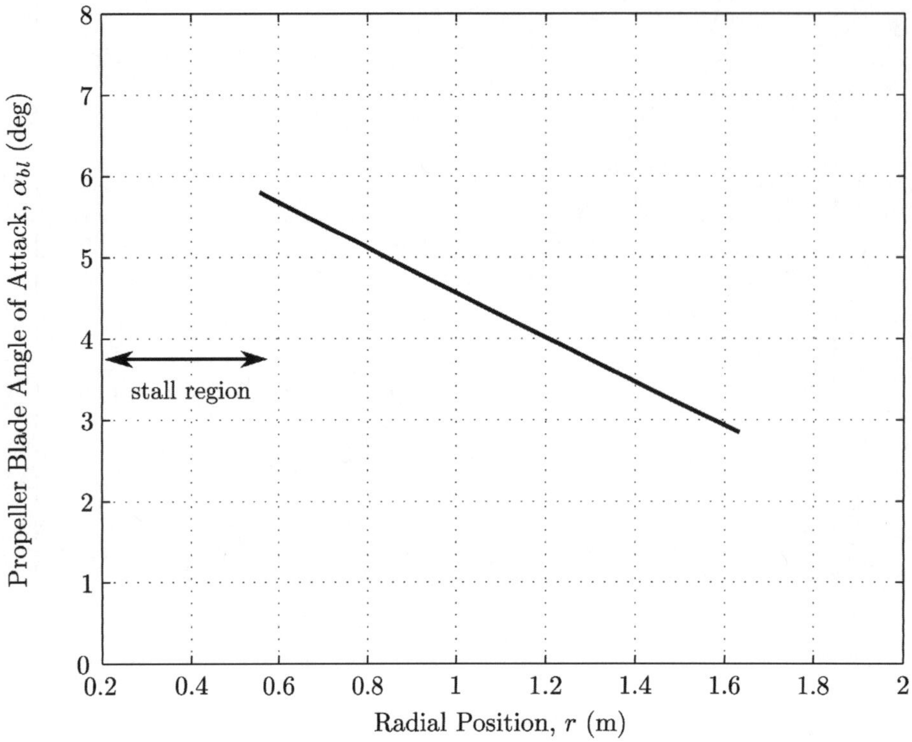

Fig. 5. Calculated Propeller Twist Distribution for the Cheyenne Compound

occurring at the inner sections of the propeller blade. The angle of attacks begin to lower towards the outer portion of the blade due to the effects of compressibility.

The result is a propeller planform with a high level of twist, which promotes favourable angles of attack across the blade at the cruise condition. The mean chord calculated by the propeller design algorithm is 0.258 m, giving a propeller solidity of 0.19. The propulsive efficiency at cruise is 0.85, which is close to maximum achievable efficiency. It is interesting to note than when compressibility effects are not modelled, by assuming linear aerodynamics, then the maximum propulsive efficiency achievable is 0.95. This is clearly unreasonable and reinforces the need to model compressibility effects to capture the performance of the propeller.

INVERSE SIMULATION ALGORITHM

The inverse simulation algorithm used is the so called integration method. This method is well documented in the literature (Refs. 4, 5, 22), therefore only a brief description is provided in the main text. The integration method uses numerical integration and conventional simulation to calculate the controls required to move a vehicle through a desired trajectory (Ref. 4). Essentially, inverse simulation reverses the process of traditional simulation, Figure 6. The first step is to calculate the control angles that trim the aircraft for the given starting airspeed. Generally, a helicopter can be in trimmed flight when climbing, descending or flying with a lateral velocity (sideslip). However in this current work the trimmed state corresponds to steady level flight with the body accelerations and the attitude rates equal to zero. Concerning the UH-60A helicopter, the trim algorithm calculates the four control angles, roll and pitch angles which result in zero translational and angular accelerations acting at the aircraft's center of gravity. Essentially, there are six trim targets which are

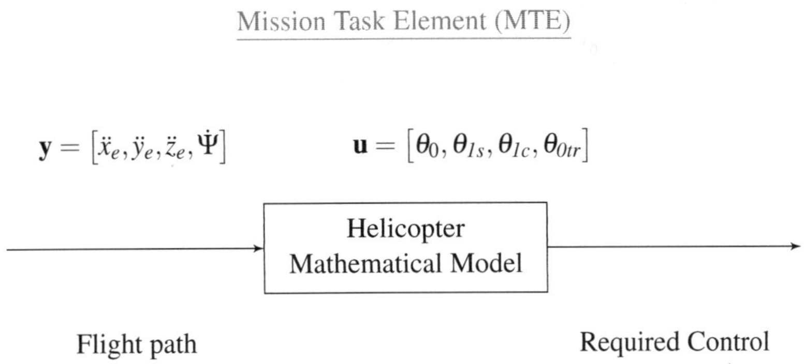

Fig. 6. Schematic showing the process of Inverse Simulation

$$X - mg\sin\Theta \tag{10}$$
$$Y + mg\cos\Theta\sin\Phi \tag{11}$$
$$Z + mg\cos\Theta\cos\Phi \tag{12}$$
$$L = 0 \tag{13}$$
$$M = 0 \tag{14}$$
$$N = 0 \tag{15}$$

which correspond to the condition of steady level flight. However, the introduction of an extra control to the Cheyenne compound requires a slightly different approach to trim these vehicles. The Cheyenne configuration features five controls and a description of the controls is shown in Table 2. There are two approaches taken to determine the control angles required to trim the compound helicopter. The first method trims the compound helicopter in low speed flight, whereas the second calculates the trim controls in forward flight. In low speed flight, an additional trim constraint is added, to make a total of seven trim targets. This additional constraint is to match the propulsive thrust of the propeller(s), depending on the compound configuration, to a prescribed thrust value. This approach minimises the power of the propeller(s) in low speed flight. For the Cheyenne configuration, the additional constraint is

$$T_{prop} = X \tag{16}$$

This constraint (or trim target) is added to the six trim targets, Equations (10) - (15), making a total of seven trim targets to match the seven unknowns. The term X in Equation (16) is a prescribed thrust value which is selected to minimise the power of the propellers in low speed flight.

In forward flight the trim algorithm is switched to a different approach. The pitch attitude of the compound helicopter is fixed to a zero pitch attitude, $\Theta = 0$ deg, in forward flight. The choice to constraint the compound helicopters to a level fuselage is an attempt to minimise the drag of the fuselage and to ensure that the aircraft's wing operates close to its maximum lift-to-drag ratio. The angle of incidence of the wing is selected to be 7 deg, for the compound vehicle, which gives a lift coefficient of the wing, C_L, of approximately 0.6 in forward flight. This results in the wing of the compound providing a significant amount of lifting force whilst retaining an adequate stall margin, at low drag. An indirect consequence of this trim approach is that the two propellers provide the majority of propulsive force to overcome the airframe drag in high speed flight.

The next step, after the calculation of the trim control angles, is to define the manoeuvre. The manoeuvre is discretised into a series of discrete time points, t_k, by specifying the time step. Subsequently, the manoeuvre can be determined with a matrix, $y_{des}(t_k)$ representing the flight path of the manoeuvre. The manoeuvre can be defined by polynomials that satisfy the requirements of the particular manoeuvre (Ref. 23), with the mathematical modelling

Control Symbol	Control Name	Control Axis
θ_0	Main rotor collective	Heave control
θ_{1s}	Main rotor longitudinal cyclic	Pitch control
θ_{1c}	Main rotor lateral cyclic	Roll control
θ_{prop}	Propeller pitch	Propeller thrust control
θ_{tr}	Tail-rotor pitch	Yaw control

Table 2. Controls for the Cheyenne Compound Configuration

of the Depart and Abort manoeuvre detailed later. Starting from the trimmed condition, $\boldsymbol{u}_e$ is the initial guess to calculate the control vector, $\boldsymbol{u}$, to force the helicopter to the position of the next time point. A Newton-Raphson technique is used to calculate the control vector to force the vehicle to the next time point to match the desired flight path defined by $\boldsymbol{y}_{des}(t_k)$. After convergence, this numerical technique moves onto the next time point and repeats the process. The end result is the control activity required throughout the manoeuvre.

MANOEUVRE MODELLING

To successfully implement inverse simulation, the trajectory the aircraft is to follow is required. In the early inverse simulation algorithms, the output vector $\boldsymbol{y}_{des}$, which describes the trajectory, was defined as follows:

$$\boldsymbol{y}_{des} = \begin{bmatrix} x_e & y_e & z_e & \Psi \text{ or } \beta \end{bmatrix}^{\mathrm{T}} \tag{17}$$

where x_e, y_e and z_e are the flight path co-ordinates and the additional constraint is either the heading or sideslip angle. As previously stated, the integration method is a flexible and robust approach to solve the inverse problem. However, the development of this approach provided new numerical instabilities, such as the control activity featuring "low-amplitude, high frequency oscillations superimposed on the low frequency waveform" as discussed by Hess, Gao and Wang (Ref. 24). Originally, the desired output vector was composed of the flight path co-ordinates with an additional constraint, either the heading or sideslip angles, as seen in Equation (17). However, Rutherford and Thomson showed that altering the output vector to be compose of inertial accelerations removed the high frequency oscillations of the control activity (Ref. 25). Therefore, modifying the output vector to the following

$$\boldsymbol{y}_{\mathrm{des}} = \begin{bmatrix} \ddot{x}_e & \ddot{y}_e & \ddot{z}_e & \dot{\Psi} \text{ or } \dot{\beta} \end{bmatrix}^{\mathrm{T}} \tag{18}$$

was shown to attenuate these oscillations (Ref. 5). Consequently, in this current work the output vector consists of the aircraft accelerations and heading or sideslip rate. The accelerations are determined by the differentiation of the velocities in Earth axes. The velocities, in Earth Axes, are conveniently described by the trajectory angles (glideslope and track angles) as seen in Figure 7. The velocities, are therefore given by

$$\dot{x}_e = V \cos\gamma \cos\chi \tag{19}$$
$$\dot{y}_e = V \cos\gamma \sin\chi \tag{20}$$
$$\dot{z}_e = -V \sin\gamma \tag{21}$$

leading to the following accelerations

$$\ddot{x}_e = \dot{V} \cos\gamma \cos\chi - V\dot{\gamma} \sin\gamma \cos\chi - V\dot{\chi} \cos\gamma \sin\chi \tag{22}$$
$$\ddot{y}_e = \dot{V} \cos\gamma \sin\chi - V\dot{\gamma} \sin\gamma \sin\chi + V\dot{\chi} \cos\gamma \cos\chi \tag{23}$$
$$\ddot{z}_e = -\dot{V} \sin\gamma - V\dot{\gamma} \cos\gamma \tag{24}$$

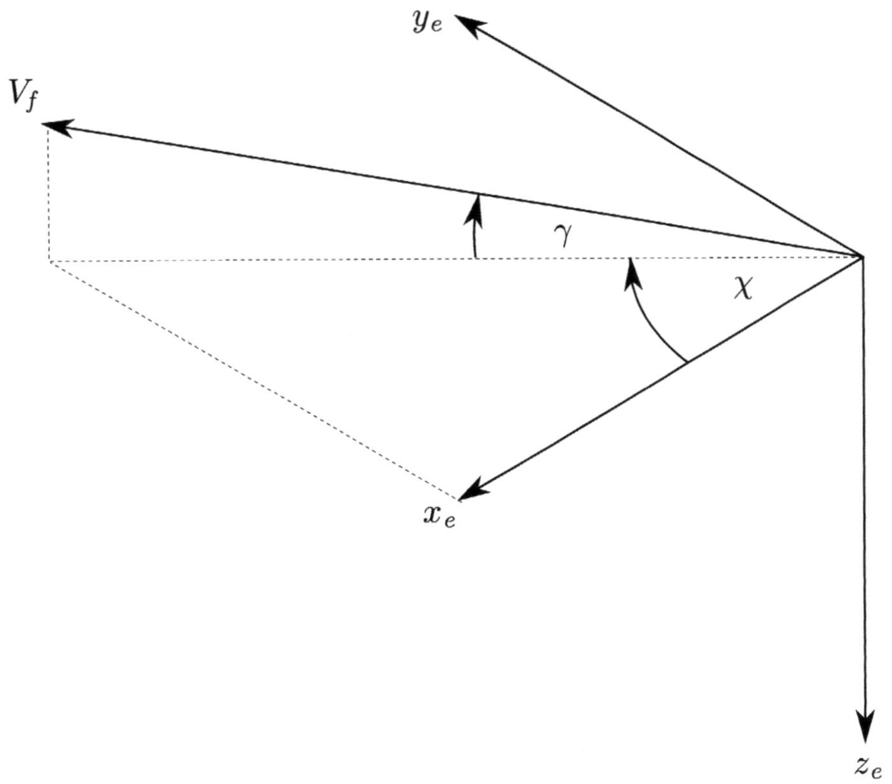

Fig. 7. Diagram showing the definition of the glideslope and track angles

Therefore if γ, χ and V are all defined and their time derivatives calculated, then the accelerations can be determined, hence forming the output vector.

Alternatively, the accelerations in Earth axes may be known from flight test data. This is for the case for the Depart and Abort manoeuvre. The flightpath was obtained from flight tests conducted the US Army's research centre at Moffet Field (Ref. 26). The helicopter's position throughout the manoeuvre were taken and discretised using the appropriate time step. Thereafter, the aircraft's position was differentiated twice to obtain the helicopter's accelerations in the Earth axes set.

Justification of the Manoeuvre Selection

In theory, inverse simulation can be used to calculate the control activity required to fly any manoeuvre, with the assumption that an appropriate output vector, Equation (18), can be formed. There are a wide selection of manoeuvres which helicopters are required to perform routinely. In this study, the manoeuvre selection is based on two factors. Firstly, it seems prudent to select a manoeuvre where the addition of compounding is likely to have a significant influence. The second factor in the selection of the manoeuvre is the fidelity of the helicopter mathematical model. As discussed in the Section , the level 1 simulation model (which is the focus of the study) is not suitable for flight regimes where the main rotor aerodynamics become highly nonlinear due to compressibility and reverse flow effects. Hence, the selected manoeuvre should begin and end when the helicopter is operating well within its Operational Flight Envelope (OFE). As a consequence, the use of a level 1 model prohibits the use of the inverse simulation algorithm in high speed flight. The use of the Level 2 simulation model would be required to assess the compound helicopters at the edge of their perceived flight envelope.

To satisfy the these conditions, the study focuses on the Depart and Abort test manoeuvre, which is described in the ADS-33 specification (Ref. 27). An additional benefit of the Depart and Abort manoeuvre is that the research team have access to flight test corresponding to this particular manoeuvre. The main aim of the Depart and Abort

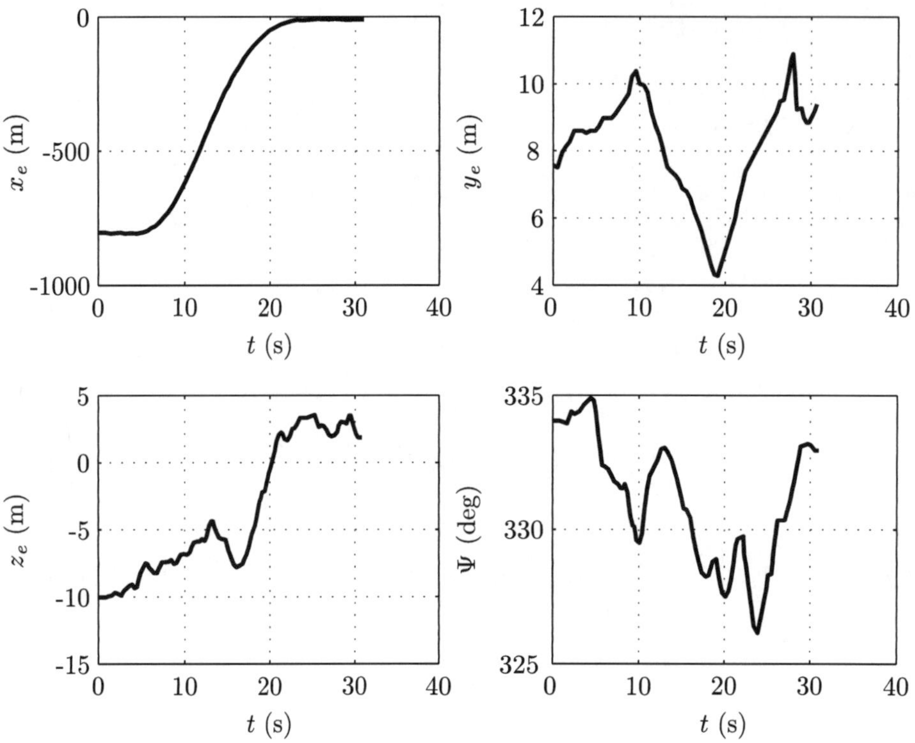

Fig. 8. Flight path of the Depart and Abort Mission

manoeuvre is to test the aircraft's pitch and heave axis handling qualities (Ref. 27). One undesirable effect of the conventional helicopter, is the large pitch excursions required to accelerate the aircraft. As the main rotor is the sole source of propulsion for a conventional helicopter, the rotor is required to tilt forward so that a component of the rotor thrust vector provides a propulsive force. This tilt of the main rotor also comes with the penalty of pitching the aircraft's nose down. However, the introduction of thrust compounding offers the potential of reducing the pitch excursion, as the propeller can provide the required propulsive force. There are also questions relating to the design of the propeller. For example, how much propulsive force can thrust compounding provide during the Depart and Abort test manoeuvre? Also, what is the best control strategy to fly this manoeuvre? Overall, there are a variety of research questions, which this study attempts to answer.

Although only the Depart and Abort manoeuvre is considered in this study, it is possible that this work could be extended to investigate the performance and handling qualities of compound helicopters in high speed manoeuvres. The natural consequence of compounding the helicopter is the expansion of the flight envelope, exposing the helicopter to unfamiliar flight conditions. Consequently, the compound helicopter will be able to perform high speed level turns or high-speed pull ups. These types of manoeuvres should be considered in future studies. An investigation of these types of manoeuvres would need to be supported by a level 2 rotor model, which is capable of modelling nonlinear aerodynamics such as retreating blade stall and compressibility effects. Lateral-directional manoeuvres, such as the slalom, could also be considered in future studies. The addition of propeller(s) to the helicopter's airframe will generate gyroscopic moments when the helicopter is pitching, rolling or turning. These moments could degrade the handling qualities of the vehicle and this type of study could provide some interesting results. Another issue is that compound helicopters typically do not have a conventional tail-rotor, meaning that the yaw control of the vehicle is provided by some other mechanism. Therefore, a manoeuvre like the slalom could potentially be used to compare the effectiveness of the devices which are used for yaw control.

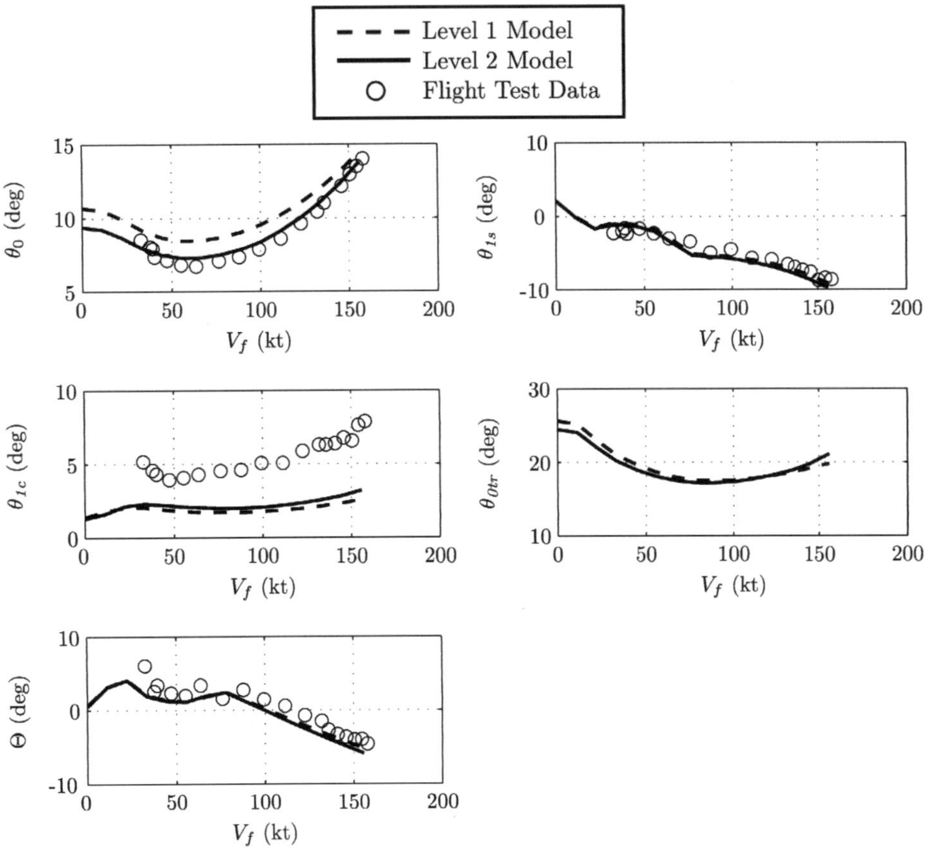

Fig. 9. UH-60A Trim Results using the Levels 1 and 2 Mathematical Models

RESULTS

Validation of the UH-60A Model

One of the first tasks is to compare the conventional helicopter simulation model with flight test data, in order to gain confidence of the mathematical model. Fortunately, there is data available from UH-60A flight tests which can be used for model validation (Refs. 28, 29). The UH-60A simulation model is trimmed and compared with results from the flight test designated FLT-85 (Ref. 28). The results are shown in Figure 9, and the agreement between the simulation and flight test data is generally very good. Also shown is the comparison between the two simulation models, the levels 1 and 2 models, which were introduced in Section . The level 1 model over-predicts the collective angle across the flight envelope. For the level 2 model (the model which includes nonlinear aerodynamics), the comparison between the collective angle and the corresponding flight test results is excellent. It is important to note that the results above 130 kt, where blade stalling may occur, is excellent. The level 2 does not include a dynamic stall model, but it doesn't seem to adversely influence the trim results. Despite this, it is recognised that dynamic stall is important aerodynamic phenomenon, particularly in aggressive manoeuvring flight, and it is recommended a dynamic stall model is included into the simulation model, at a later date. The prediction of the longitudinal cyclic angles, using both the levels 1 and 2 models, is excellent across the entire flight envelope. This result is particularly impressive considering the moveable tailplane of the UH-60A, with the UH-60A simulation model successfully modelling its effect to the pitch axis. The variation of the tailplane's incidence is why the longitudinal cyclic changes quite abruptly at some parts of the flight envelope. As expected, the pitch attitude results are also excellent. There is quite a large difference between the lateral cyclic results, Figure 9. This also occurs in another modelling study (Ref. 28), which uses the state of the art model, CAMRAD II. The reason for the disparity of the lateral cyclic results is unknown. Unfortunately, the data recorded in FLT-85 does not include the tail-rotor collective

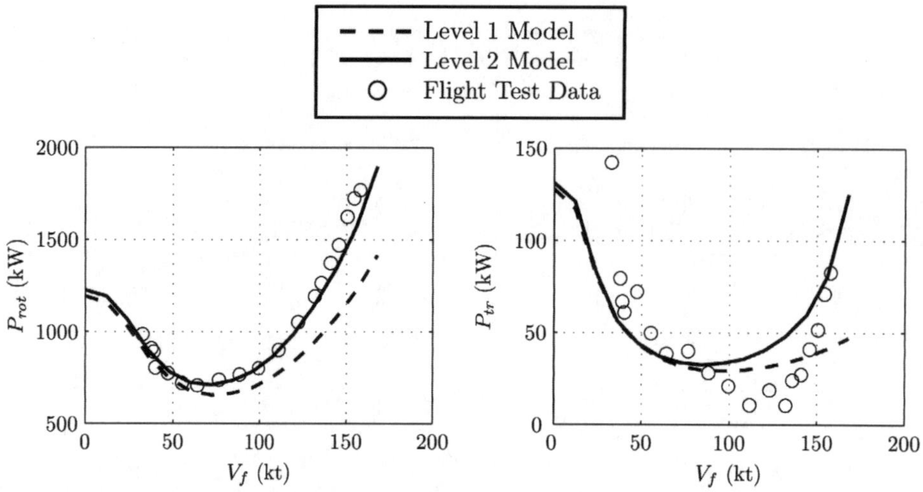

Fig. 10. UH-60A Power Predication using the Levels 1 and 2 Mathematical Models

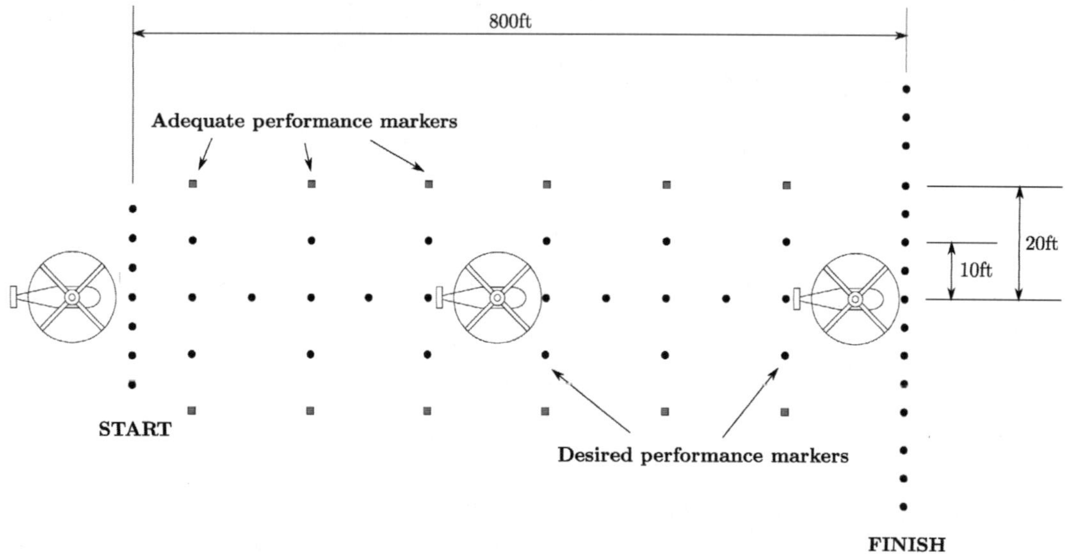

Fig. 11. Depart and Abort Test Course from ADS-33 (Ref. 27)

results.

Figure 10 compares the main rotor and tail-rotor power results with the flight test data from FLT-85. For the main rotor's power, the results of the level 2 model show excellent agreement across the whole flight envelope. In terms of the level 1 model, the results compare favourably with the flight test data, but only between the hover and 70 kt. Thereafter, the level 1 model under-predicts the main rotor power. The level 1 model assumes that the profile drag per blade element is based on a constant drag profile coefficient. As a result, this model is incapable of successfully predicting the main rotor's profile drag rise in forward flight, due to its inability to model nonlinear aerodynamics. In contrast, the level 2 model is capable of modelling compressibility effects and blade stalling, by using comprehensive look-up tables of the SC-1095 aerofoil (Refs. 8, 9). The end result here is that a level 2 model is required to successfully predict the main rotor's power in forward flight. In terms of the tail-rotor power, both models compare favourably with the flight test data until an airspeed of 70 kt. After 70 kt, both models do not predict the tail-rotor power well. These differences are also observed in the work of Yeo and Johnson (Ref. 28), and the reason for the disparity is unclear.

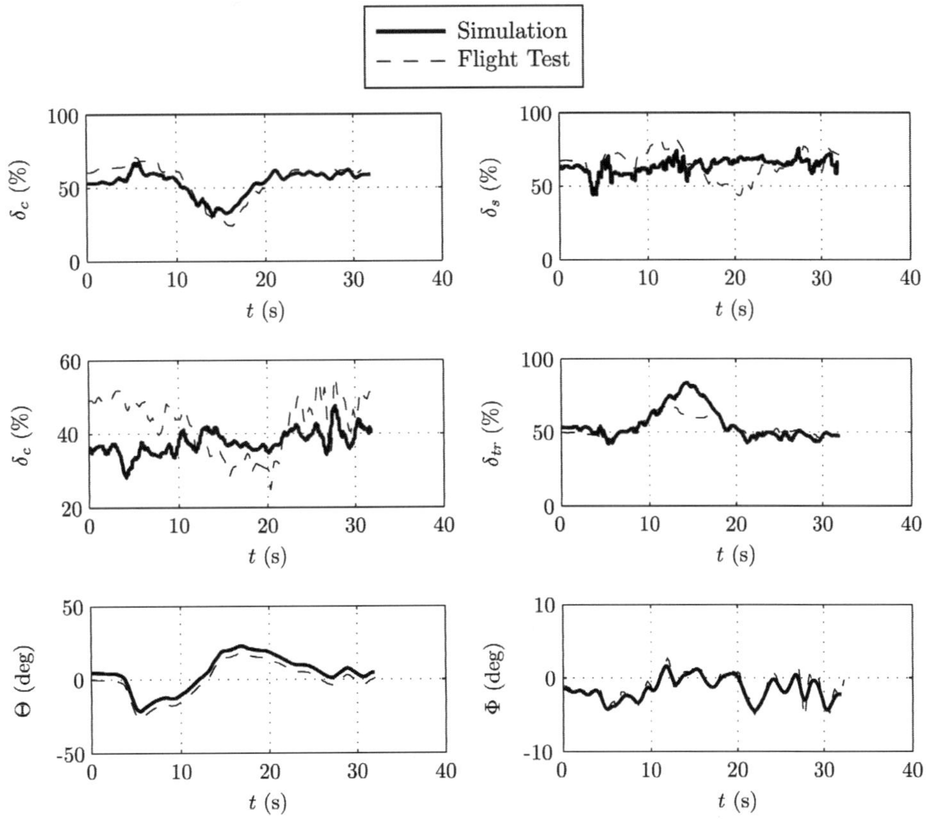

Fig. 12. Comparison between the Control Positions during the Depart and Abort Test Manoeuvre

The level 1 simulation model has been validated in steady level flight, and some issues with the model have been identified. A more rigorous test of the simulation model is to calculate the control angles in manoeuvring flight and compare them with flight test data. Fortunately, one of the benefits of inverse simulation is that it is possible to fly the simulation model through a manoeuvre which is identical to a flight test trial, and then compare time histories (Ref. 30). The manoeuvre of interest is the Depart and Abort MTE, and the corresponding test course is shown in Figure 11. On the test course, Figure 11, there are a series of cones placed to define the adequate and desired performance standards. In this task the helicopter starts in the hover and is required to accelerate forward to an airspeed of 40 kt, then return to a stabilised hover at the end of the 800 ft long course. The manoeuvre is only concluded when the pilot manages to place the helicopter's cockpit within 20 ft of the end point of the test course. Figure 12 shows the level 1 simulation model results of the UH-60A Blackhawk helicopter, flying the Depart and Abort MTE. The results from the simulation are compared with flight data from US Army flight trials (Ref. 26). The US Army flight trial results where obtained in clear weather conditions during the summer of 1999 (Ref. 26). The results are encouraging, particularly the agreement with the main rotor collective and the pedal inputs. The longitudinal and lateral stick movements are fairly well predicted, although there are some discrepancies between the simulation and flight test results. The agreement between the pitch and roll attitudes of the aircraft are excellent. Overall, the agreement is very good, highlighting that the level 1 simulation model, coupled with the inverse simulation algorithm, can be used to successfully predict the behaviour of the helicopter during this low speed test manoeuvre.

DEPART AND ABORT RESULTS

The process to calculate the conventional helicopter's controls through a manoeuvre was described earlier, and a schematic of the inverse simulation algorithm is reproduced in Figure 13, to support the discussion. To apply

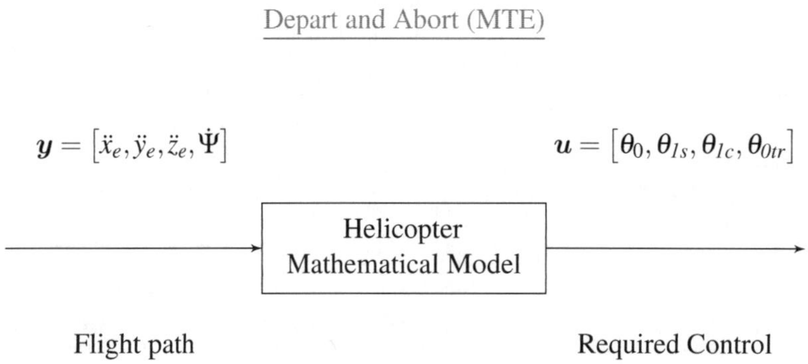

Fig. 13. Schematic showing the process to calculate the control of the Conventional Helicopter

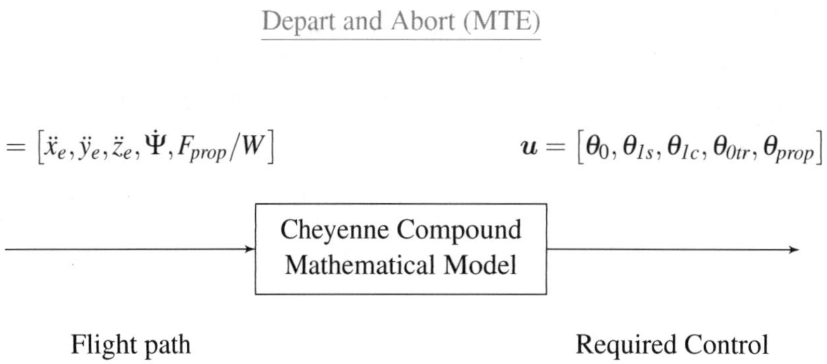

Fig. 14. Schematic showing the process to calculate the controls of the Cheyenne configuration

the inverse simulation algorithm successfully, the amount of unknowns must match the number of inputs. This occurs with the conventional helicopter, as there are four inputs, which define the manoeuvre, and four control angles to be calculated, Figure 13. Since the compound helicopter feature five controls, an additional constraint is required, so that there are five inputs which match the five unknown controls. A logical question which arises with the compound helicopter flying the Depart and Abort manoeuvre is how much of the propulsive force is provided by the propellers. Conveniently, the fifth constraint for this manoeuvre is selected to be the ratio between the propeller's propulsive force (i.e. the force in the x_e direction) and the aircraft's weight. This allows the simulation to regulate how much propulsive force thrust compounding is required to create throughout the manoeuvre. For the Cheyenne compound, the inverse solution is shown in Figure 14. There are five inputs into the inverse simulation algorithm: three accelerations, the heading angle and the ratio between the propulsive force the propeller is required to generate and the vehicle's weight. The algorithm subsequently calculates the five controls for this vehicle to fly this manoeuvre in the appropriate manner, Figure 14.

It is likely that in the acceleration phase of the Depart and Abort manoeuvre that the pilot or control system would adopt a control strategy which would actively use thrust compounding to provide some portion of the propulsive force throughout the manoeuvre. However, it is unknown how much thrust the propellers are able to provide. Secondly, it is not known what is the optimum control strategy to minimise the power required by each compound configuration. As a result, the amount of propulsive thrust provided by the propellers is varied in multiple simulation runs to determine how sensitive the power required is to the control strategy adopted. The amount of thrust required by the propellers is gradually increased to the point where the propellers are incapable of producing the required thrust. Figure 15 shows how the ratio between the propulsive force, required by thrust compounding, and the aircraft's weight changes with each simulation run. The propulsive force ratio (F_{prop}/W) is formed by creating an 8th order polynomial as follows

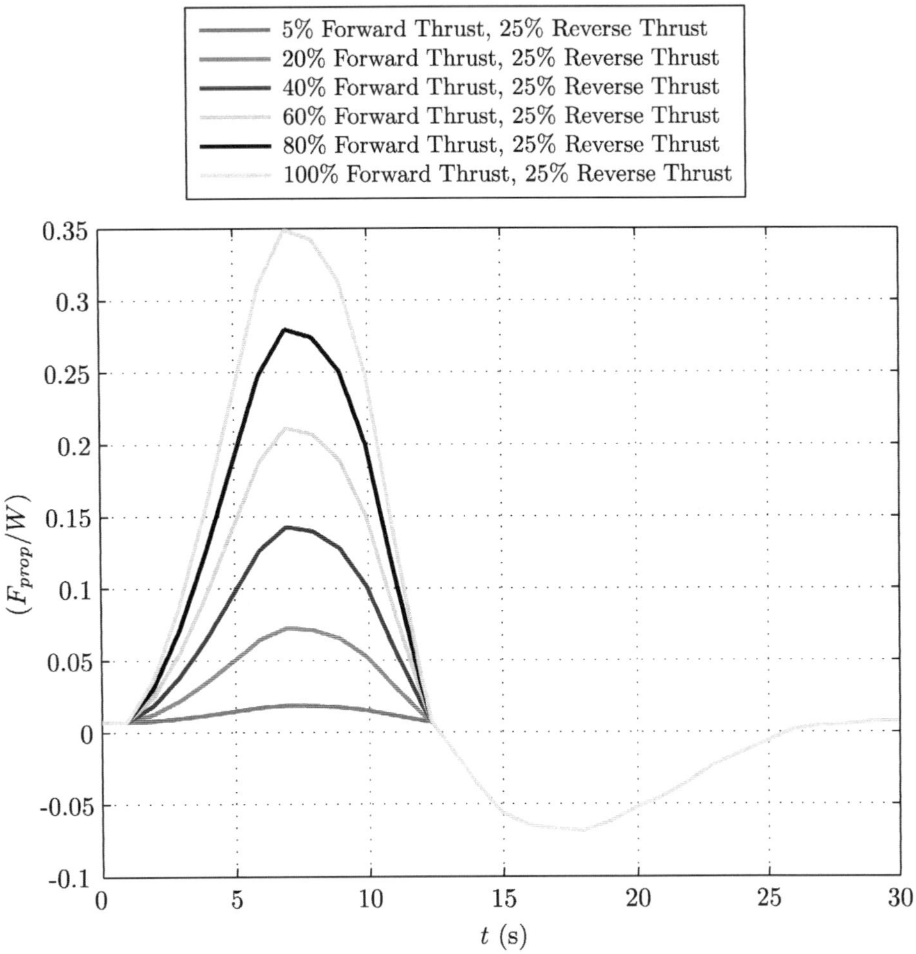

Fig. 15. Propulsive Force Provided by Thrust Compounding

$$\frac{F_{prop}}{W} = c_0 t^8 + c_1 t^7 \dots c_7 t + c_8 \tag{25}$$

where c_0, c_1, etc. are coefficients. The coefficients of Equation (25) can be easily altered to promote greater propulsive thrust throughout the manoeuvre. Returning to Figure 15, the first case (shown in red) is the minimal propulsive thrust case where the thrust compounding is not required to provide any meaningful propulsive thrust during the forward acceleration part of the manoeuvre. In this scenario, the compound helicopter flies the first part of the manoeuvre with traditional like helicopter manoeuvring. For the second case (which is shown in magenta) thrust compounding is required to provide more of the total propulsive thrust at the point of maximum forward acceleration, which occurs at 7 s. The amount of the propulsive thrust generated by thrust compounding is gradually increased until some aerodynamic restriction is met, Figure 15. An additional advantage of thrust compounding is the possibility of using the propellers to decelerate the aircraft by providing reverse thrust. It is fair to assume that the design of the propellers would be optimised for high speed flight and not for reverse thrust purposes where the propellers operate in the windmill brake state. Hence it is assumed that the propellers provide 25% of the entire reverse force required to decelerate the vehicle at the time of peak deceleration. For every case, the amount of reverse thrust create by thrust compounding is 25%. Note that no comprehensive investigation was conducted to investigate the optimal amount of reverse thrust, and this is a recommendation for future work.

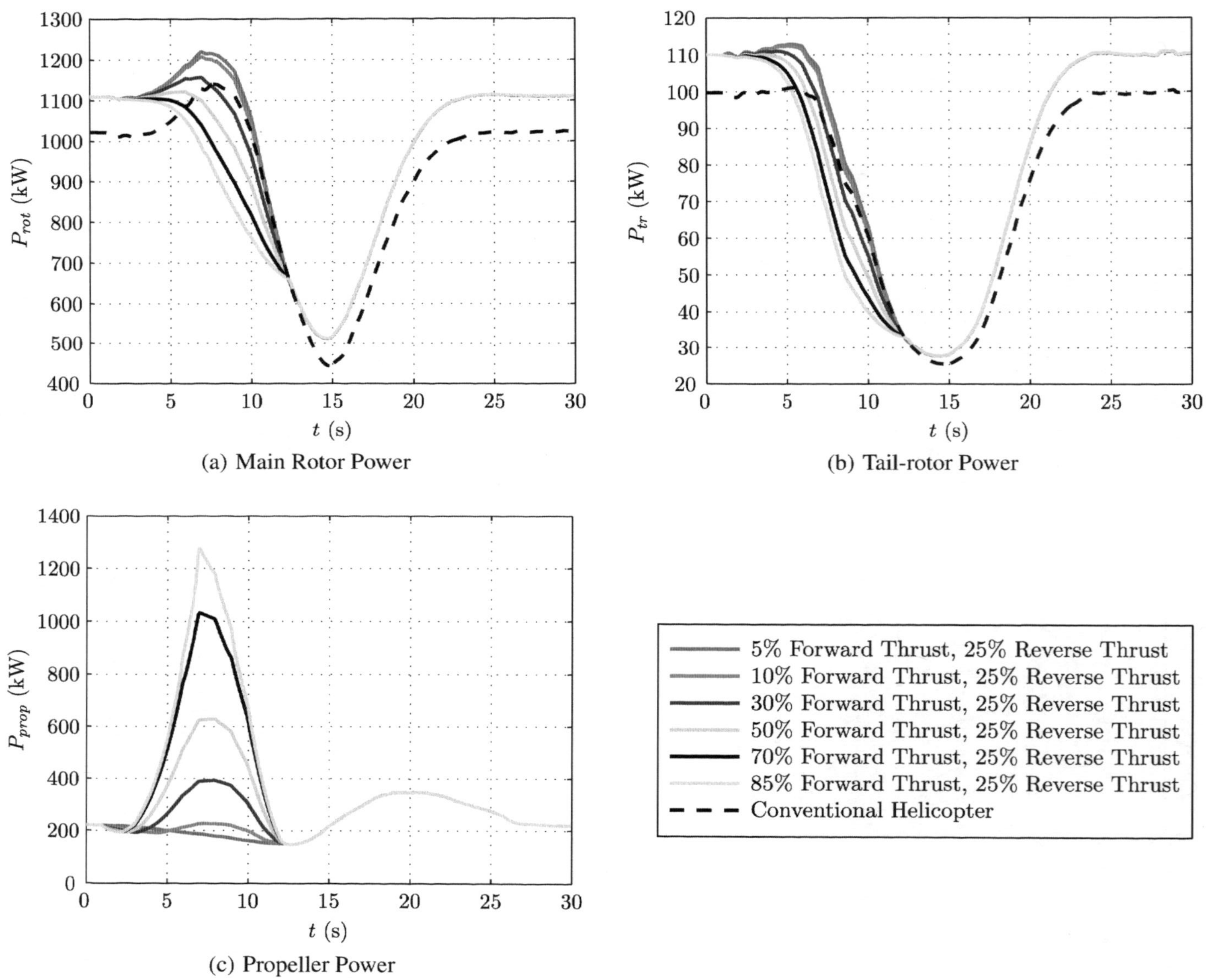

(a) Main Rotor Power

(b) Tail-rotor Power

(c) Propeller Power

Fig. 16. Power Requirements of the Cheyenne Configuration

Cheyenne Configuration Results

Figure 16 (shown on the previous page) presents the power of the main rotor, tail-rotor and pusher propeller during the Depart and Abort manoeuvre for the various test cases considered. The results shown in Figure 16 correspond to the Cheyenne configuration with a pusher propeller of solidity 0.19. This solidity value was determined early in the study by focusing on the cruise condition of the Cheyenne configuration. It is predicted that the original propeller design of Cheyenne configuration is incapable of producing enough thrust to provide the entire propulsive force during the forward acceleration part of the manoeuvre. The result is that the propeller of the Cheyenne can only produce 85% of the total propulsive thrust to accelerate the aircraft forward, before the propeller suffers from blade stalling. Consequently, Figure 16 presents the test cases where the pusher propeller of the Cheyenne provides 5, 10, 30, 50, 70 and 85% of the propulsive thrust during the first part of the manoeuvre. The main rotor power is shown in Figure 16(a) and follows the expected trend. When comparing the 85% case of the Cheyenne compound, to the conventional helicopter, greater rotor power is required by the conventional helicopter between 6-12 s. This is due to the pusher propeller significantly offloading the main rotor of its propulsive duties. Although this is favourable, in terms of reducing the main rotor's power, there is a large increase to the power required by the pusher propeller, Figure 16(c). In the deceleration part of the manoeuvre, between 12-30 s, there is a slight rise to the amount of power required by the pusher. This is due to the pusher propeller providing some of amount reverse thrust to slow the airspeed of the aircraft. For the Cheyenne compound, the power of the tail-rotor is fairly insensitive to the different test cases. As expected, as the pusher propeller is required to provide more thrust to accelerate the

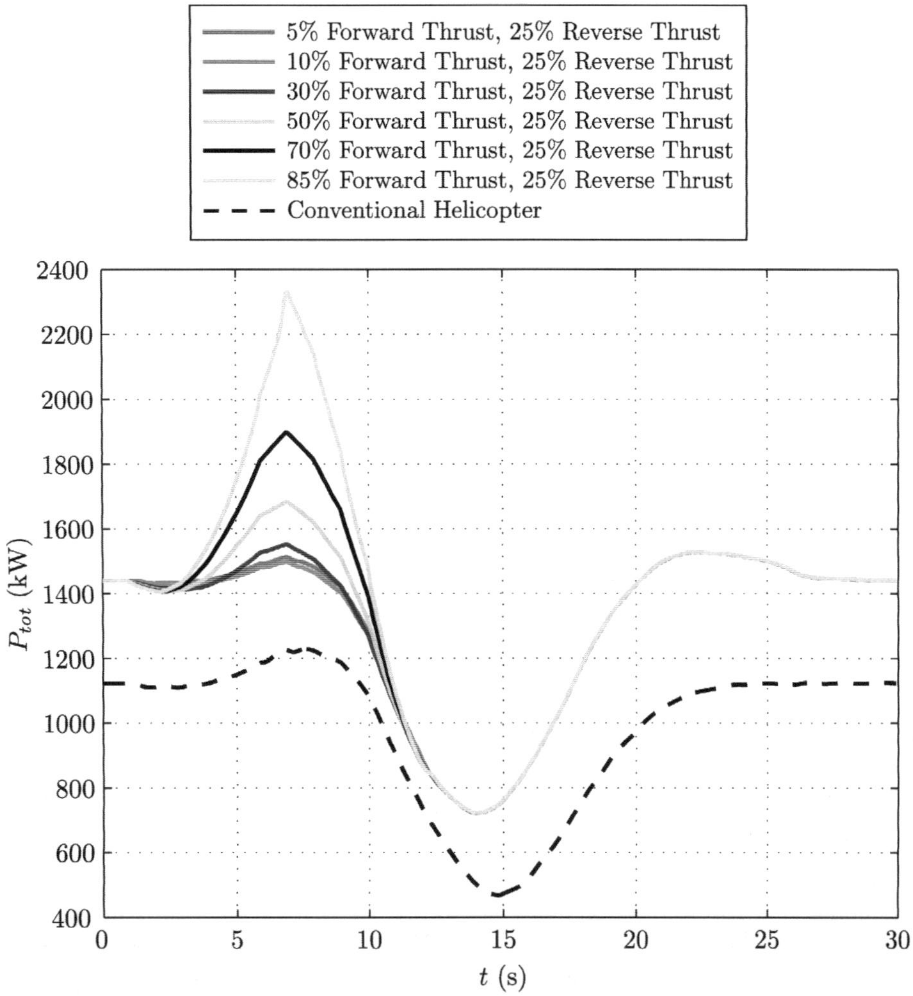

Fig. 17. Total Power Requirements of the Cheyenne Compound with a Propeller Solidity of 0.19

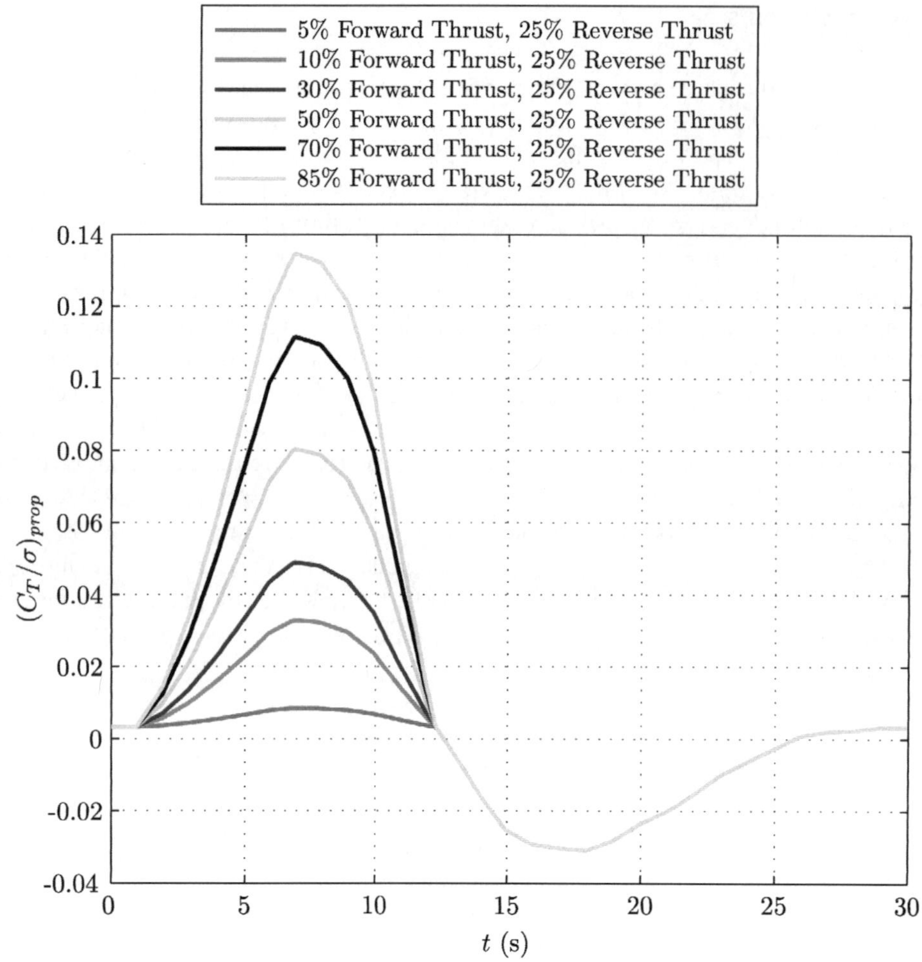

Fig. 18. Blade Loading of the Cheyenne Compound's Propeller using a Solidity of 0.19

helicopter forward, the propeller power increases, Figure 16(c).

Figure 17 presents the total power required by the Cheyenne compound with a propeller solidity of 0.19. For this particular configuration, the results show that the most efficient way to accelerate the aircraft forward, from the hover, is to minimise the amount of propeller thrust required by the pusher propeller. For the 5, 10 and 30% cases, there is little difference between the amount of power required to accelerate the Cheyenne forward. However, as the pusher propeller is required to contribute to more than 30% of the propulsive force, there is a general rise of the amount of power required between 3-11 s. This result further emphasises the ability of the conventional main rotor to efficiently produce a propulsive force in low speed flight. This is clearly demonstrated in Figure 17 by comparing the total power required by the conventional helicopter and the Cheyenne compound's 80% case, at 7 s. For the conventional helicopter, the total power required is 1207 kW, compared to the 2310 kW required by the Cheyenne configuration. This is a significant difference and suggests that if the Cheyenne compound were to fly the manoeuvre in this manner then it would require more installed power, than the conventional helicopter. Another general point worth highlighting is that the power required by the Cheyenne configuration is greater than that of the conventional throughout the entire manoeuvre, regardless of the test case considered.

As mentioned previously, with the current design, the pusher propeller is only capable of producing 85% of the total propulsive force during the forward acceleration part of the manoeuvre. Figure 18 presents the blade loading of the pusher propeller the Depart and Abort manoeuvre. The maximum blade loading value of the pusher propeller is achieved at the point of maximum forward acceleration, at approximately 7 s. As the aircraft begins to decelerate, the pusher propeller provides some reverse thrust which can be seen with the negative blade loading values after 12 s. The propeller of the Cheyenne compound's pusher propeller reaches its maximum thrust capability when

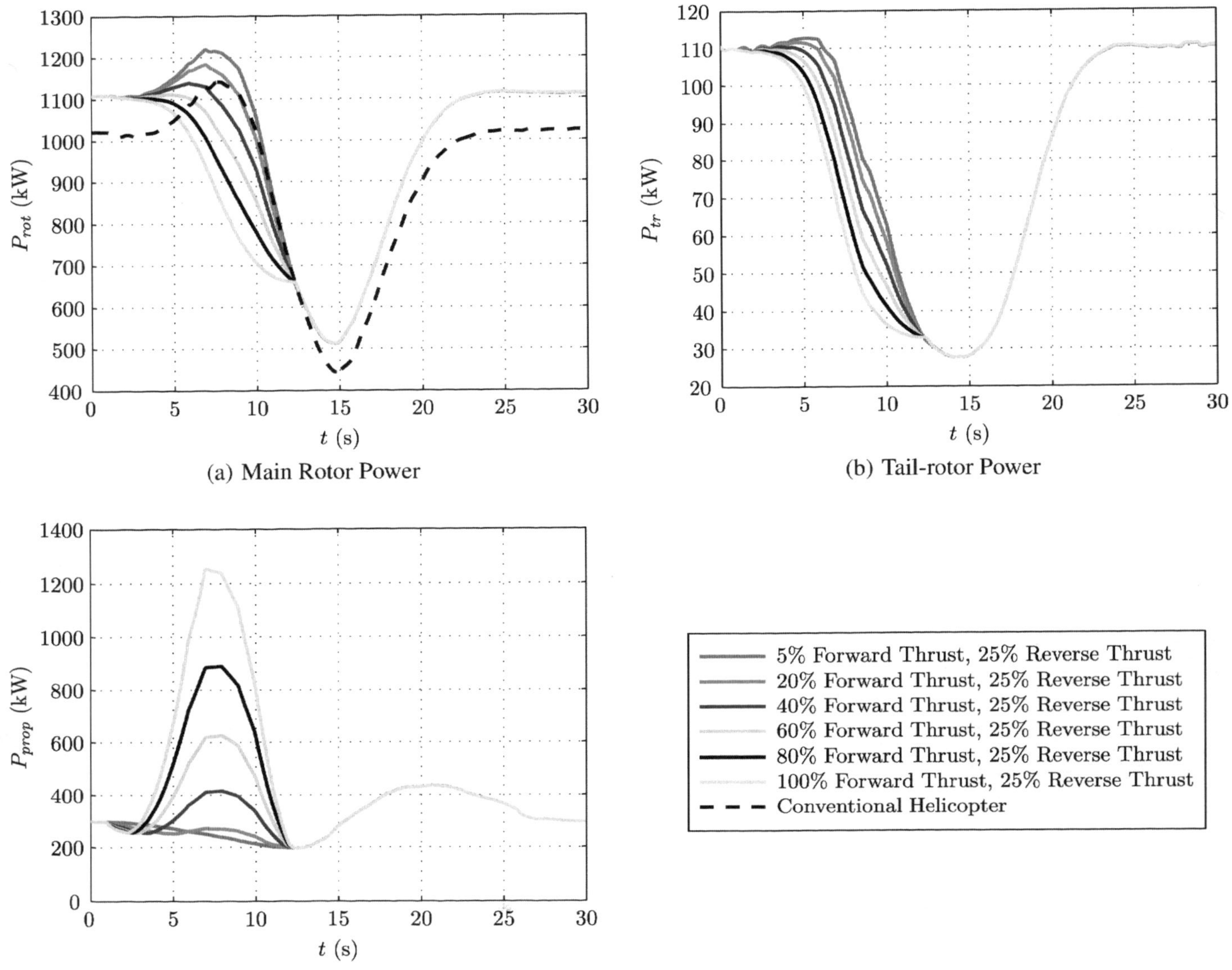

(a) Main Rotor Power

(b) Tail-rotor Power

(c) Propeller Power

Fig. 19. Power Requirements of the Cheyenne Configuration with a Propeller Blade Solidity of 0.25

$(C_T/\sigma)_{prop} \approx 0.14$. Above this blade loading value, the angles of attack about the propeller disc become large, leading to blade stalling. For the pusher propeller to provide the entire propulsive force during the first part of the manoeuvre, requires an increase of blade solidity.

The solidity of the Cheyenne compound's pusher propeller is increased to 0.25, and the corresponding rotor power, tail-rotor power and pusher propeller power results are shown in Figure 19 (previous page). The results shown in Figure 19 are generally very similar to the results presented in Figure 16. The only major difference between the results is the change in the required power by the pusher propeller, due to its greater solidity. By increasing the propeller's solidity to 0.25, then it is predicted that the pusher propeller can completely offload the main rotor of its propulsive duties, during the first part of the manoeuvre. This is confirmed by inspecting the blade loading values of the pusher propeller throughout the manoeuvre, as shown in Figure 20. The increased solidity avoids the pusher propeller reaching the blade stalling limit, which occurs at approximately 0.14. At the 100% case, the maximum power of the pusher propeller reaches 1270 kW, which is greater than the total power required to hover the conventional helicopter at mean sea level. Although the pusher propeller is capable of providing all of the propulsive force to accelerate the aircraft forward, a significant amount of power is required. In low speed flight, the main rotor is clearly more efficient than the pusher propeller in creating propulsive thrust. This result is important for the purposes of designing the flight control system. The compound helicopter is likely to feature a digital fly-by-wire system due to benefits of reducing weight as well as enhancing the handling qualities of the aircraft. For a pilot to fly the Depart and Abort manoeuvre, with minimal workload,

would require an attitude/velocity hold response, as discussed by Fletcher et.al, in the fly-by-wire upgrade to the UH-60 (Ref. 31). In this mode, a forward stick motion increases airspeed and the pilot is able to hold a constant

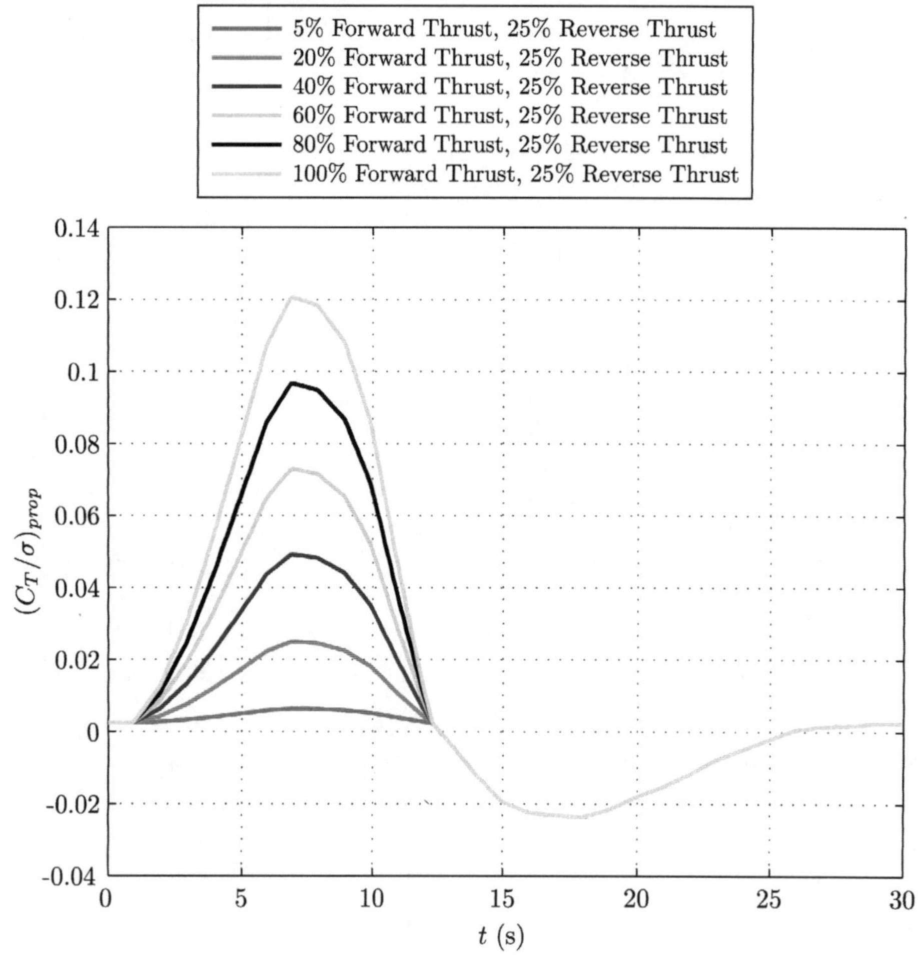

Fig. 20. Total Power Requirements of the Cheyenne Compound with Propeller with a Solidity of 0.25

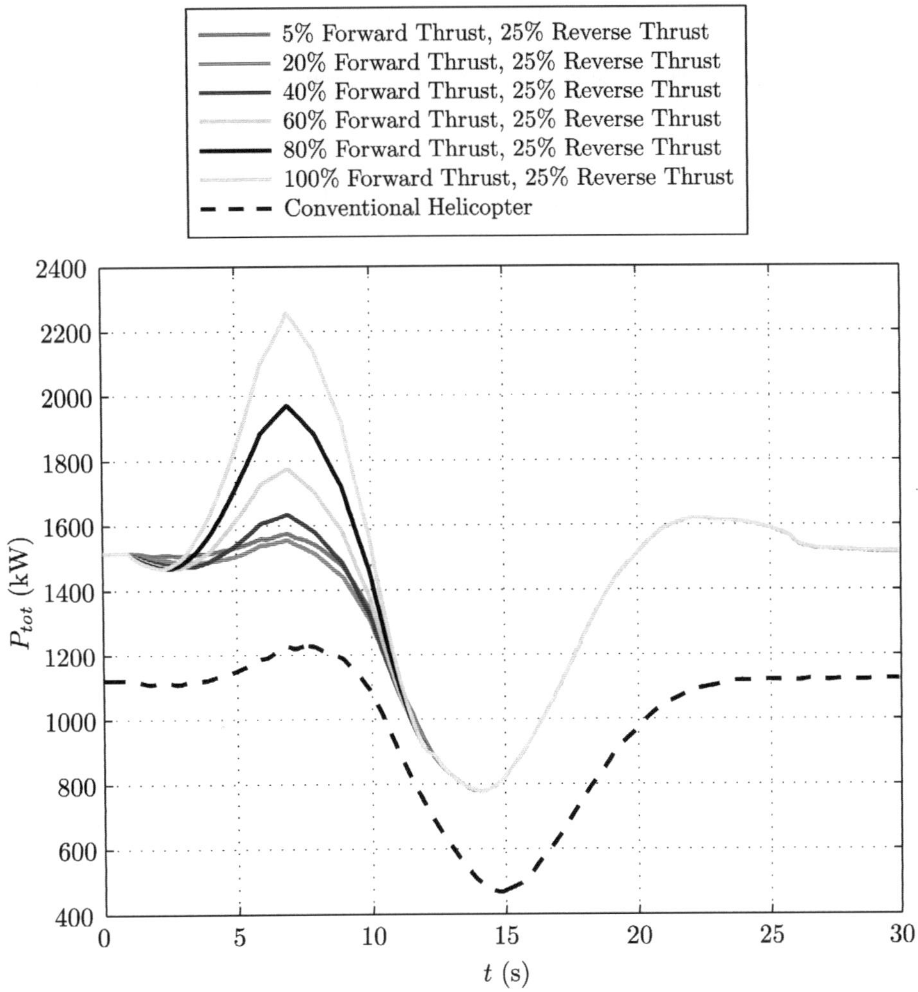

Fig. 21. Total Power Requirements of the Cheyenne Compound with a Propeller Solidity of 0.25

airspeed by re-centring the stick. This control mode would be suitable for the compound helicopter. However, various simulation studies would be required to develop suitable control laws to mix the longitudinal cyclic and propeller pitch controls, so that the vehicle can achieve a smooth acceleration or deceleration by the use of a single control. The results presented in this study show that in low speed flight, the design of the control laws may need to be tailored so that the propeller and the main rotor share the propulsive duties, to avoid flight conditions where the power required is not excessive.

Figure 21 presents the total power required by the Cheyenne configuration, using a propeller solidity of 0.25. Again the results suggest that the Cheyenne compound is able to accelerate forward, from the hover, most efficiently when the main rotor provides the majority of the propulsive thrust. Due to thrust compounding, the pilot of a compound helicopter will have the option to accelerate or decelerate the aircraft by using the propellers or the main rotor. The description of the Depart and Abort test manoeuvre, which will apply to compound helicopters, may have to allow the pilot to fly the manoeuvre in a variety of ways. The first way could be using traditional helicopter manoeuvring, by fully exploiting the main rotor to accelerate and decelerate the aircraft. The second way could involve the pilot sharing the propulsive duties with both the main rotor and the propeller. Lastly, the pilot could fly the manoeuvre with the propeller completely divorcing the main rotor of its propulsive duties. By flying the manoeuvre in a variety of ways, the pilot will be able to check the response of the helicopter for the different control modes. In addition, the pilot will be able to check of the flight control system successfully harmonises the main rotor and propeller controls together.

CONCLUSIONS

The broad aim of this study was to assess the performance of compound helicopters flying the Depart and Abort ADS-33 Mission Task Element. The main conclusions from the paper, are as follows:

1. The propeller design of the compound helicopter is likely to feature highly twisted blades for good propulsive efficiency in forward flight.
2. The level 1 model is capable of predicting the trim control angles of the conventional helicopter, with reasonable accuracy, across the flight envelope. However, a level 2 model is required to accurately predict the main rotor's power requirements in trimmed forward flight.
3. The level 1 is able to successfully predict the pilot controls of the UH-60A helicopter, flying the Depart and Abort MTE.
4. The simulation model predicts the maximum blade loading of the propeller during the Depart and Abort manoeuvre is approximately 0.14.
5. When using a propeller solidity of 0.19 the propeller of the Cheyenne compound helicopter is only capable of providing 85% of the entire total propulsive force during the Depart and Abort manoeuvre.
6. The propeller of the Cheyenne compound helicopter is required to have a solidity of 0.25, so that the propeller can provide the entire total propulsive force during the Depart and Abort manoeuvre.
7. The total power required by the Cheyenne compound helicopter, when flying the Depart and Abort MTE, is minimised when the propeller provides a small portion of the required propulsive thrust. The main rotor is clearly more efficient than the propeller in creating propulsive thrust in low speed flight.

REFERENCES

[1]Hirschberg, M., "Joint Multi-Role Moves Forward," *Vertiflite*, Vol. 60, (1), 2014, pp. 24–26.

[2]Padfield, G., *Helicopter Flight Dynamics: the Theory and Application of Flying Qualities and Simulation Modelling*, Blackwell Publishing, second edition, 2007.

[3]Ferguson, K., "Towards a better understanding of the flight mechanics of compound helicopter configurations," PhD Thesis, University of Glasgow, 2015.

[4]Thomson, D. and Bradley, R., "Inverse simulation as a tool for flight dynamics research - Principles and applications," *Progress in Aerospace Sciences*, Vol. 42, (3), may 2006, pp. 174–210.
doi: 10.1016/j.paerosci.2006.07.002

[5]Rutherford, S., *Simulation Techniques for the Study of the Manoeuvring of Advanced Rotorcraft Configurations*, Ph{.}{d} thesis, University of Glasgow, 1997.

[6]Yeo, H. and Johnson, W., "Optimum Design of a Compound Helicopter," *AIAA Journal of Aircraft*, Vol. 46, (4), 2009.
doi: 10.2514/1.40101

[7]Prouty, R., *Helicopter Performance, Stability, and Control*, Robert E. Krieger Publishing Company, Inc., reprint edition, 1990.

[8]Howlett, J., "UH-60A Black Hawk Engineering Simulation Program: Volume 1 - Mathematical Model," NASA CR-10626, 1984.

[9]Howlett, J., "UH-60A Black Hawk Engineering Simulation Program: Volume 2 - Background Report," NACA CR-166310, 1988.

[10]Orchard, M. and Newman, S., "The fundamental configuration and design of the compound helcopter," *Proceedings of the Institution of Mechanical Engineers, Part G: Journal of Aerospace Engineering*, Vol. 217, (6), 2003, pp. 297–315.
doi: 10.1243/095441003772538570

[11]Keys, C., "Performance Prediction of Helicopters," *Rotary-wing Aerodynamics*, edited by W. Stepniewski, Chap. Volume II, 1979.

[12]Frandenburgh, E. and Segel, R., "Model and full scale compound helicopter research," Helicopter Society 21st Annual, 1965.

[13]Torres, M., "A Wing on the SA.341 Gazelle Helicopter and its Effects," *Vertica*, Vol. 1, (1), 1976, pp. 67–73.

[14]Sipe, O. and Gorenberg, N., "Effect of Mach Number, Reynolds Number and Thickness Ratio on the Aerodynamic Characteristics of NACA 63A-Series Airfoil Sections," USATRECOM TR 65-28, 1965.

[15]Van Dyke, M., "High-Speed Subsonic Characteristics of 16 NACA Six Series Airfoil Sections," NACA TN 2670, 1952.

[16]Adkins, C. and Liebeck, R., "Design of Optimum Propellers," *Journal of Propulsion and Power*, Vol. 10, (5), 1994, pp. 676 – 682.

[17]Wald, Q., "The aerodynamics of propellers," *Progress in Aerospace Sciences*, Vol. 42, (2), 2006, pp. 85–128.
doi: 10.1016/j.paerosci.2006.04.001

[18]Ormiston, R., "Low-Disk Loading Compound Rotorcraft for High-Speed and Aerodynamic Efficiency," International Powered Lift Conference 2010, 2010.

[19]Sforza, P., *Commercial Airplane Design Principles*, Elsevier Aerospace Engineering Series, London, first edition, 2014.

[20]Raymer, D., *Aircraft Design: A Conceptual Approach*, AIAA Education Series, Virginia, fourth edition, 2006.

[21]Marinus, B. and Poppe, J., "Data and design models for military turbo-propeller aircraft," *Aerospace Science and Technology*, Vol. 41, (February), 2015, pp. 63–80.
doi: 10.1016/j.ast.2014.12.009

[22]Hess, R. and Gao, C., "A Generalized Algorithm for Inverse Simulation Applied to Helicopter Maneuvering Flight," *Journal of the American Helicopter Society*, Vol. 38, (4), 1993, pp. 3 – 15.
doi: 10.4050/JAHS.38.3

[23]Thomson, D. and Bradley, R., "Mathematical Definition of Helicopter Maneuvers," *Journal of the American Helicopter Society*, Vol. 4, (1), 1997, pp. 307 – 309.
doi: 10.4050/JAHS.42.307

[24]Hess, R., Gao, C., and Wang, S., "Generalized technique for inverse simulation applied to aircraft maneuvers," *AIAA Journal of Guidance, Control and Dynamics*, Vol. 14, (5), 1991, pp. 920–926.
doi: 10.2514/3.20732

[25]Rutherford, S. and Thomson, D., "Helicopter inverse simulation incorporating an individual blade rotor model," *AIAA Journal of Aircraft*, Vol. 34, (5), 1997.

[26]Blanken, C., Cicolani, L., Sullivan, C., and Arterburn, D., "Evaluation of Aeronautical Design Standard - 33 Using a UH-60A Black Hawk," American Helicopter Society 56th Annual Forum, 2000.

[27]Anon., "Handling qualities requirements for military rotorcraft," Aeronautical design standard ADS-33E-PRF, United States Army Aviation and Troop Command, 2000.

[28]Yeo, H., Bousman, W., and Johnson, W., "Performance analysis of a utility helicopter with standard and advanced rotors," *Journal of the American Helicopter Society*, Vol. 49, (3), 2004, pp. 250–270.
doi: 10.4050/JAHS/49.250

[29]Bousman, W. and Kufeld, R., "UH-60A Airloads Catalog," NASA/TM-2005-212827, 2005.

[30]Bradley, R., Padfield, G., Murray-Smith, D., and Thomson, D., "Validation of helicopter mathematical models," *Transactions of the Institute of Measurement and Control*, Vol. 12, (186), 1990.
doi: 10.1177/014233129001200405

[31]Fletcher, J., Lusardi, J., Mansur, M., Robinson, D., Arterburn, D., Cherepinsky, I., Driscoll, J., Morse, C., and Kalinowksi, K., "UH-60M Upgrade Fly-By-Wire Flight Control Risk Reduction using the RASCAL JUH-60A In-Flight Simulator," American Helicopter Society 64th Annual Forum, 2008.

Development and Simulation Testing of High-Speed Tracking MTEs for Rotorcraft Handling Qualities Evaluation

Cody E. Fegely
Senior Engineer
Sikorsky Aircraft Corporation
Stratford, CT

Hong Xin
Senior Staff Engineer
Sikorsky Aircraft Corporation
Fort Worth, TX

William C. Fell
Principal Experimental Test Pilot
Sikorsky Aircraft Corporation
West Palm Beach, FL

David H. Klyde
VP & Technical Director, Engineering Services
System Technology, Inc.
Hawthorne, CA

Chris L. Blanken
Vehicle Management & Control TA Lead
ADD, U.S. Army RDECOM
Moffett Field, CA

ABSTRACT

The Vertical Lift Consortium (VLC) project, "Rotorcraft Handling Qualities Requirements for Future Configurations and Missions" is a U.S. Government and industry co-funded research project intended to develop mission task elements (MTEs) to ensure desired handling qualities of high-speed rotorcraft. These new high-speed tasks must fit into the mission-oriented approach already established by ADS-33E-PRF. The work presented in this paper summarized the development and simulation testing of two new tracking MTEs. These tracking MTEs, sum-of-sines (SOS) and discrete, were modified versions of similar tasks flown by fixed-wing aircraft for decades. Four aircraft configurations were used to evaluate the ability of the sum-of-sines and discrete tracking MTEs to identify known handling qualities deficiencies. The results of the evaluation showed the sum-of-sines tracking MTE aligned well with the predicted and assigned levels of handling qualities obtained from ADS-33E-PRF, while the discrete tracking MTE did not. The HQRs obtained from the SOS tracking MTE were well within ±1 HQR of the Depart/Abort and Slalom MTE for each configuration evaluated.

INTRODUCTION

The current industry-standard military design specification for rotorcraft handling qualities, ADS-33E-PRF (Ref.1), covers the hover/low-speed and forward flight regimes for the conventional rotorcraft configuration, for example, single main rotor plus tail rotor. ADS-33E-PRF requirements seem well-tuned for the current military fleet, mainly comprised of 1960-80's rotorcraft technology. Future Vertical Lift (FVL) is about the next revolutionary period in vertical lift. FVL requirements are likely to result in unconventional configurations capable of achieving much higher forward flight speeds than the conventional rotorcraft. FVL is comprised of several rotorcraft classes (light, medium, heavy) with each rotorcraft capable of performing a variety of roles (attack, reconnaissance, assault). Revisions to ADS-33E-PRF may be needed to ensure sufficient handling qualities throughout the operational flight envelope for new rotorcraft being developed as a result of the requirements from FVL.

The Vertical Lift Consortium (VLC) project, "Rotorcraft Handling Qualities Requirements for Future Configurations and Missions" is a U.S. Government and industry co-funded research project designed to collaboratively investigate and develop a comprehensive update to the mission task elements (MTEs) which are required for evaluating different rotorcraft configurations with respect to the FVL requirements. The project, led by Sikorsky Aircraft Corporation as the prime contractor, is a collaborative effort of industry and academia. Under this project, efforts have been made by Sikorsky and Systems Technology, Inc. (STI) to develop a set of tracking MTEs for high-speed, precision tasks. A piloted simulation study was performed using a high-fidelity model of an aircraft utilizing X2™ technology. The development and simulation testing of these new tracking MTEs are presented in this paper.

Presented at the AHS Rotorcraft Handling Qualities Technical Meeting, Huntsville, Alabama, USA, February 22-23, 2017.

BACKGROUND

For the military rotorcraft, handling qualities are specified using a highly successful mission-oriented approach (Ref.1). The foundation of the mission-oriented approach (Ref.2) is that requirements are based on realistic MTEs, not general flight phases. In the mission-oriented approach, references to aircraft size are removed. For example, if a particular mission requires a high level of aggressiveness and precision, it should not matter if the aircraft proposed for that mission is small or large. Only the mission requirements should set the handling qualities. It is understood that in some cases this may lead to unreasonable demands on very large airplanes. In such an event, a new MTE may be required that better represents the mission of that aircraft.

One of the most significant features of the mission-oriented approach is the inclusion of the demonstration maneuvers as an integral part of the standard. The addition of demonstration maneuvers allows for two separate methods for assessing the levels of handling qualities: *Predicted Levels* based on flying qualities parameters and *Assigned Levels* based on flight test maneuvers. The predicted handling qualities level consists of comparisons with quantitative boundaries of flying qualities parameters. The assigned handling qualities level consists of performing a set of well-defined flight test maneuvers using a team of three test pilots evaluators. These pilots assign handling qualities ratings (HQRs) to the aircraft for each maneuver based upon the Cooper-Harper scale (Ref.3).

In a fixed-wing, mission-oriented specification, the flight phase categories are defined in terms of the level of precision and aggressiveness which are required of the pilot. Four MTE categories are defined as follows (Ref.4):

- *Non-Precision, Non-Aggressive*: Non-precision tasks that require only a moderate amount of closed-loop control fall into this category. Examples include cruise flight tasks such as heading and altitude changes.
- *Non-Precision, Aggressive*: Category includes large amplitude maneuvering tasks that emphasize control power over precise dynamic response. Example tasks for this category include gross acquisition of air or ground targets.
- *Precision, Non-Aggressive*: Category includes tasks where considerable precision is required, but without the aggressive control activity. Example tasks include aerial refueling as the receptor, precision offset landing, and pitch/roll attitude captures.
- *Precision, Aggressive*: Category includes precision tasks, where extremely crisp and predictable response to control inputs is required. Examples include air-to-air and air-to-ground fine tracking.

The new high-speed MTEs must fit into the mission-oriented approach already established in ADS-33E-PRF. If the above categories were to be used in ADS-33, a method would need to be established to determine a precision task from a non-precision task, as well as, an aggressive task from a non-aggressive task. The task of developing new mission-oriented high-speed MTEs began with an extensive literature review of a variety of sources ranging from the current rotorcraft (Ref.1) and fixed-wing (Ref.5) handling qualities design standards to industry and government written papers presented at various AIAA and AHS events. Ref.6 discussed substantial lessons learned since the ADS-33E-PRF was published and suggested to continue development of a handling qualities database for the FVL family. Ref.7 indicated based on recent U.S. Army rotorcraft operations, the current ADS-33E-PRF may not have suitable coverage of aggressive air combat maneuvers.

The major focus of the literature review was on fixed-wing handling qualities requirements and demonstration maneuvers that were relevant to potential high-speed rotorcraft missions. Based upon the literature review, two types of precision, aggressive MTEs were proposed for development for rotorcraft usage: sum-of-sines and discrete tracking.

MTE DEVELOPMENT

The range of human pilot control possibilities as part of a pilot-vehicle system (PVS) comprises of three fundamental modes of dynamic behavior (Ref.8). These correspond to the following pilot control behavior patterns:

- Closed-loop (compensatory): pilot control governs outputs conditioned by PVS errors
- Combined open-loop and closed-loop (pursuit): pilot control is dependent on PVS errors plus known or induced input commands and/or system outputs
- Open-loop (precognitive): highly-skilled and practiced pilot programmed control outputs to command the aircraft to a new state

In both compensatory and pursuit operations, the pilot is highly interactive with the aircraft on a continuous basis – as in a precision, aggressive MTE, whereas in precognitive operations the pilot's actions are essentially discrete and independent of the aircraft's response during the time period of the pilot's programmed command. In compensatory pursuit processes, the pilot is part of a feedback control process, with all of the advantages (continuous reduction of errors) and disadvantages (possible instabilities) associated with this kind of control. Many precision, aggressive MTEs are compensatory in nature, that is, the pilot is compensating for displayed errors.

Tracking a computer-generated command signal produces a highly repeatable, precise task from which measures of PVS performance can be obtained. While such measures should never be interpreted as a direct indicator of handling qualities, they are one way to monitor pilot workload and confirm that workload is as desired. In the fixed-wing community, the use of compensatory tracking tasks has been used to provide a source of information on pilot-vehicle closed-loop dynamics for decades.

Two of the most common types of tracking commands are a summation of sinewaves and the Calspan Corporation step-and-ramp (discrete tracking). The objectives of these MTEs are to:

- Evaluate handling qualities in a tight, closed-loop tracking task
- Evaluate feel system and control sensitivity characteristics

- Identify bobble or pilot-induced oscillation (PIO) tendencies

The visual cueing aids for these MTEs are an appropriate command bar and fixed pippers for desired and adequate performance. When performing these MTEs, the pilots provide inputs to minimize the displayed error between the commanded attitude and the actual attitude, within the defined performance requirements. Figure 1 shows the visual display used for these two tracking MTEs. The green command bar displays the error between the commanded attitude and the current aircraft attitude. The yellow aircraft attitude bar remains fixed, mimicking a fixed reticle. This display shows to the pilot the desired and adequate performance criteria for pitch and roll. Figure 2 is an example of how a pure pitch error is displayed to the pilot, while Figure 3 is an example of how a pure roll error is shown to the pilot. For pitch, the pilot is achieving desired performance if the green circle in the middle of the command bar is inside the white circle of the aircraft attitude bar. For roll, the pilot is obtaining desired performance if the green command bar remains inside the white triangles of the attitude bar. Similarly, adequate performance is achieved if inside the yellow shaded areas.

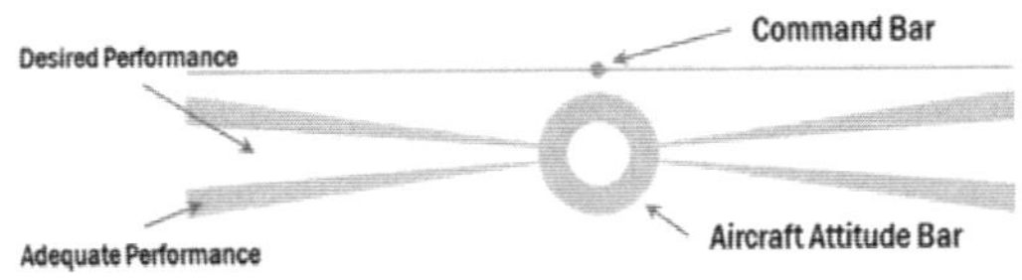

Figure 1: Visual Display for Tracking MTEs

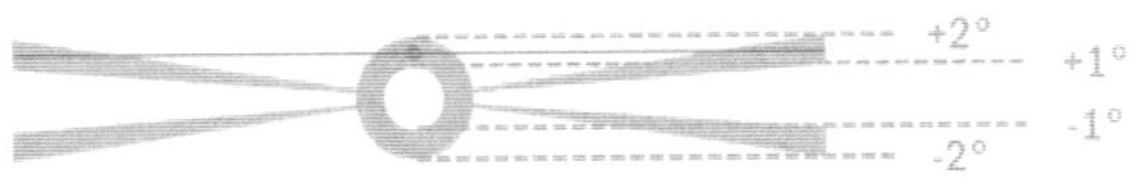

Figure 2: Example of Pitch Attitude Error Display

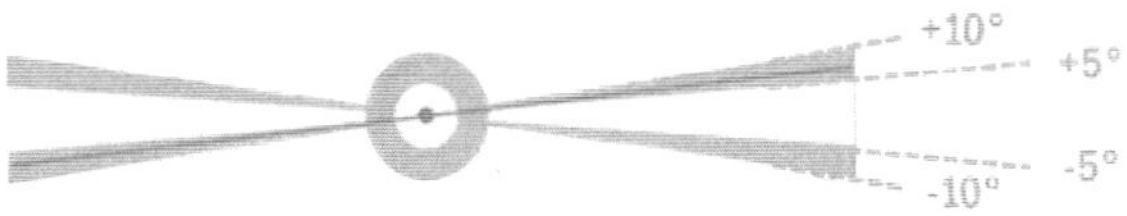

Figure 3: Example of Roll Attitude Error Display

Sum-of-Sines Tracking MTE

The form of the sum-of-sines (SOS) command is shown in equation 1.

$$X_C = \sum_{i=1}^{n} A_i \sin(\omega_i t + \phi_i) \tag{1}$$

$$\omega_n = \left(\frac{N_n}{60}\right) 2\pi \tag{2}$$

$$N_n = 5, 8, 13, 21, 34, 55, 89$$

The SOS command signal for this study utilizes seven sinewaves with frequencies that are based on a Fibonacci series. The Fibonacci series determines the cycles per run, N_n. A Fibonacci series is used since it provides equally spaced frequencies on a logarithmic scale. The command signal is defined for a 60 second scoring time run length. Thus, each sinewave frequency, ω_n (in rad/sec), is defined by equation 2. The amplitude of each of the sinewaves is determined by using a second-order Butterworth filter (equation 3). The effective bandwidth of the command signal is equal to the natural frequency of the second-order Butterworth filter which is used to determine the amplitude variation for each of the sinewaves.

$$H(s) = \frac{\omega_n{}^2}{s^2 + 2\zeta\omega_n s + \omega_n{}^2} \tag{3}$$

Figure 4 shows the magnitude response for four second-order Butterworth filters with natural frequencies of 0.50, 0.75, 1.50, and 2.50 rad/sec, respectively. The seven data points on each curve correspond to the frequencies of the seven sinewaves. The magnitude of the Butterworth filter at those specific frequencies is used to determine the amplitude variation of the sinewaves. Figure 5 describes the variation of the sinewave amplitudes for four command signal effective bandwidths. It is worth noting that the higher effective bandwidths have larger amplitudes for the sinewaves at the higher frequencies. The frequency range for the seven sinewaves is 0.5 to 9.3 rad/sec. The seven sinewave frequencies adequately cover the pilot-vehicle closed-loop frequency range.

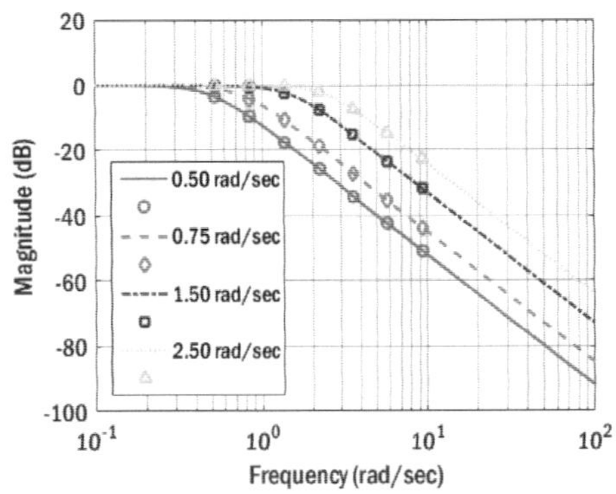

Figure 4: Magnitude for Several Second-Order Butterworth Low-Pass Filters

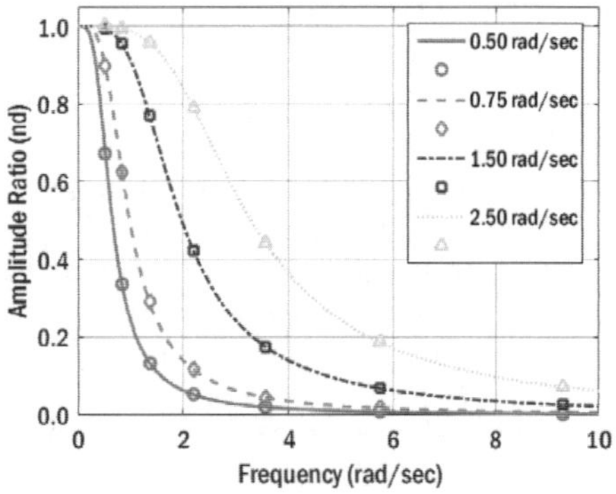

Figure 5: Amplitude Ratio of Sinewaves for Several Command Signal Bandwidths

Figure 6 shows a roll-only sum-of-sines tracking task from the first and second phase of flight testing of a fixed-wing aircraft from Ref.9. For comparison, the magnitude from a second-order Butterworth filter with natural frequencies of 0.75 and 1.50 rad/sec are included. The flight test was conducted by Calspan Corporation using the USAF/FDL NT-33A research aircraft. The command task amplitude which is used for the first phase is based on Ref.10. The sum-of-sines command inputs are designed to approximate a second-order noise process with a 0.55 rad/sec bandwidth. Thirteen sinewave components are used across a frequency range of 0.15 to 21.0 rad/sec (Ref.10). The roll SOS task in the second phase of Ref.9 has a higher bandwidth and is shaped slightly different than the first phase task. The bandwidth is on the order of 1.5 rad/sec.

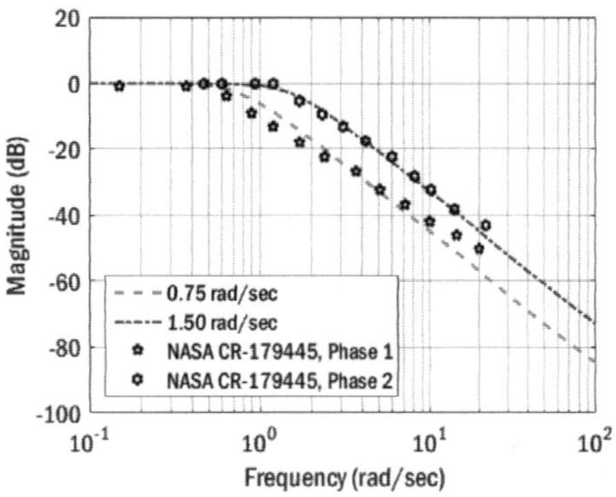

Figure 6: Calspan Roll SOS Command Task Amplitude Used for First and Second Phase Evaluations (Ref.9)

Early piloted simulator sessions were conducted to investigate the effects of command signal bandwidth and amplitude for the sum-of-sines tracking task. The SOS task was flown single axis. Handling qualities ratings were not taken during these sessions. Instead, pilot comments were taken regarding the task workload, the operational relevance, and the adequacy of the visual cues. Figure 7 depicts the root-mean-square (RMS) of the pitch tracking error for several signal magnitudes and bandwidths. As is expected, the tracking error increases with either increasing effective signal bandwidth or increasing signal magnitude. Figure 8 shows the RMS tracking error for the roll axis sum-of-sines task. The roll axis follows the same trend as shown previously for pitch. Based upon these results, the signal magnitude and bandwidth are effective knobs for modulating pilot workload and performance.

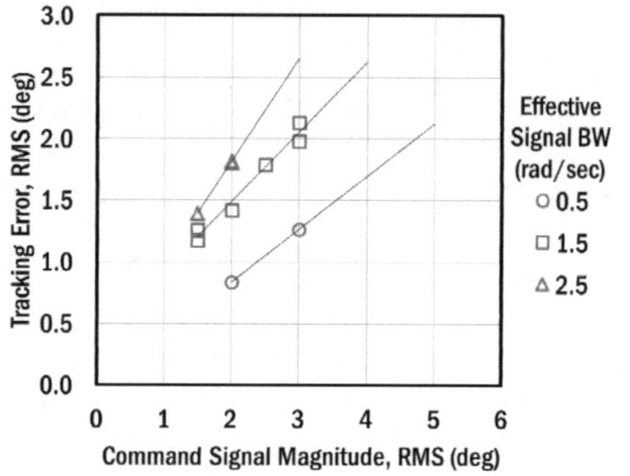

Figure 7: Pitch Axis SOS Tracking Error

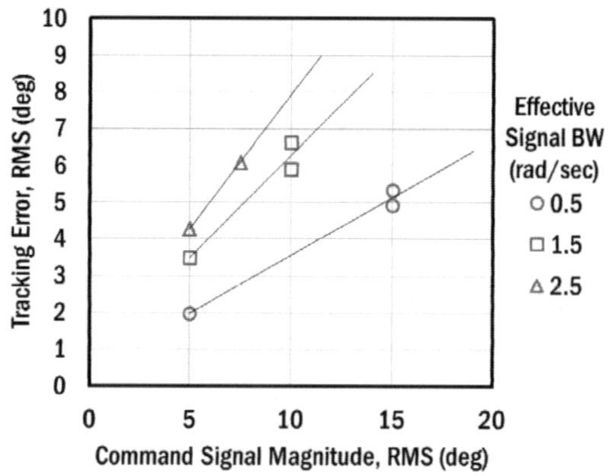

Figure 8: Roll Axis SOS Tracking Error

Discrete Tracking MTE

The most common types of discrete tracking commands are the Calspan Corporation step-and-ramp command signals (Ref.9 and Ref.10). Calspan Corporation developed and refined the step-and-ramp command signals over the years from the Neal-Smith task (Ref.11) in pitch, the Lateral Higher Order System, LATHOS, task (Ref.12) in roll, and a pitch-and-roll combination task (Ref.9).

Figure 9 and Figure 10 show the pitch and roll discrete tracking task command signals. These signals are normalized, and can be scaled as needed to target load factor and roll rate ranges.

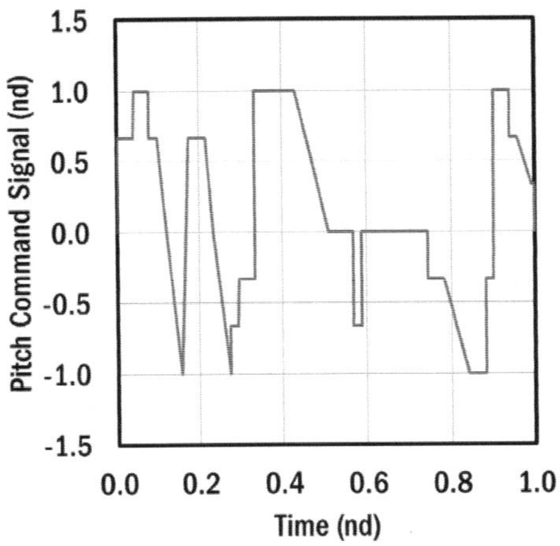

Figure 9: Pitch Discrete Tracking Task Command Signal, Normalized

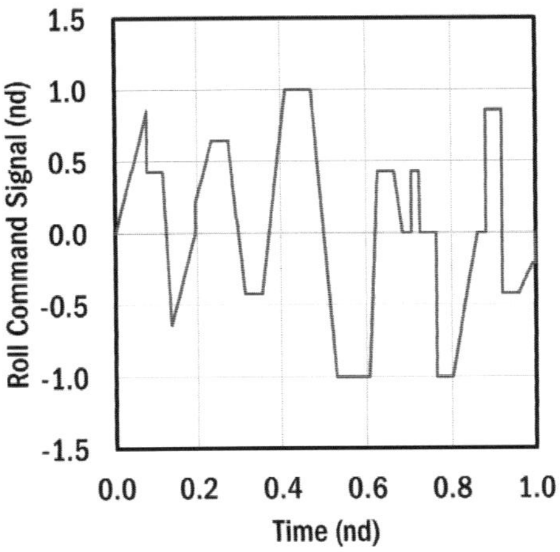

Figure 10: Roll Discrete Tracking Task Command Signal, Normalized

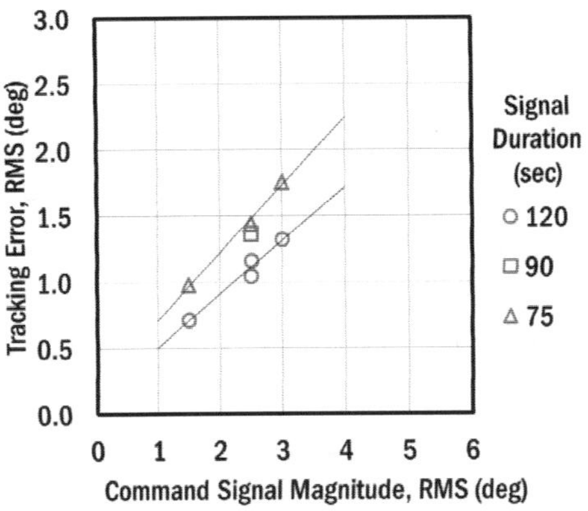

Figure 11: Pitch Axis Discrete Tracking Error

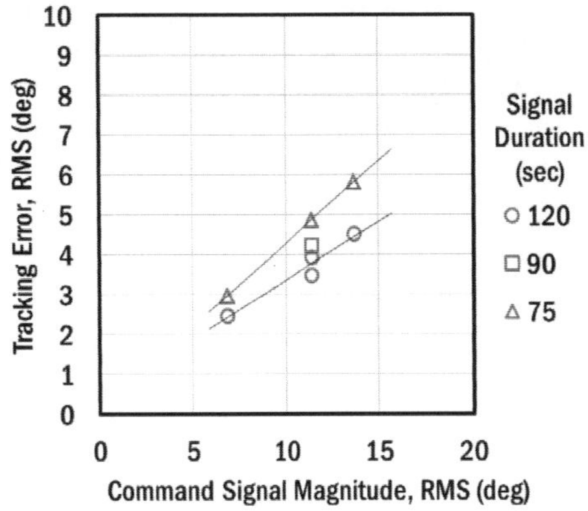

Figure 12: Roll Axis Discrete Tracking Error

Early in this program, a piloted simulator investigation was conducted to determine how signal duration and magnitude affected pilot workload and task performance. The discrete tracking task was flown dual axis (pitch and roll simultaneously). Handling qualities ratings were not taken during these sessions. Pilot comments were instead used to qualify the workload of the task, the operational relevance of the task, and adequacy of the visual cues. Figure 11 and Figure 12 show the RMS tracking error for several signal magnitudes and durations for pitch and roll, respectively. As the signal magnitude is increased and/or the signal duration is shortened, the tracking error increases for the corresponding axis. From this result, it is clear that both signal magnitude and signal duration are effective knobs for modulating pilot workload and performance.

MTE EVALUATION

Formal piloted evaluations were performed in the Sikorsky Engineering Development Simulator (EDS) shown in Figure 13. Two U.S. Army experimental test pilots and one Sikorsky experimental test pilot participated in the evaluation. Each U.S. Army pilot had limited exposure to flying any type of X2™ technology simulator prior to the evaluation. Before this study, each pilot's total X2™ simulation stick time was approximately 3 hours. In addition, the two U.S. Army evaluation pilots had no prior experience with the proposed tracking MTEs. The Sikorsky evaluation pilot had extensive experience with X2™ technology and prior exposure to the proposed MTEs.

The purpose of the evaluation is to assess the ability of the new tracking MTEs to quantify handling qualities. In order to test how well the MTEs could identify known handling qualities deficiencies, four flight control configurations are evaluated. Each configuration uses a generic GenHel simulation of an aircraft utilizing X2™ technology. All four configurations include a model following rate command/attitude hold (RCAH) response type (Ref.13) with a side-arm mounted cyclic inceptor. The command model (Ref.13 and Ref.14) is varied for each of the four configurations, while the plant, feedbacks and all other attributes remain the same. Figure 14 and Figure 15 show the predicted level of handling qualities for each of the four configurations. Configuration A is predicted to be Level 1 in both pitch and roll while the other three configurations (B, C and D) have degraded handling qualities. Each configuration in Figure 14 and Figure 15 contains the time delay that is associated with the visual system in the EDS.

The assigned level of handling qualities for each of the configurations is determined by the Depart/Abort and Slalom MTEs from Ref.1. The Depart/Abort MTE is chosen to evaluate the pitch axis because it is an aggressive MTE with a precision hover at the end. The Slalom MTE is used to evaluate the roll axis handing qualities because it is a forward flight MTE requiring precise maneuvering through a series of gates. The Accelerate/Decelerate and Sidestep MTEs would also be good choices if more aggressive but less precise maneuvering was required. Figure 16 contains the assigned handling qualities ratings for the Depart/Abort MTE. The symbol indicates the average of the three pilot ratings, while the brackets indicate the maximum and minimum rating assigned. The assigned level of handing qualities for each configuration compares well with the predicted level. The HQRs for the Slalom MTE are shown in Figure 17. Again, the assigned level of handling qualities compares favorably with the predicted levels. The HQRs from the Depart/Abort and Slalom MTE serve as a baseline for comparison to the new tracking MTEs.

The new tracking MTEs are evaluated at a forward airspeed of 120 knots. An airspeed of 120 knots was chosen as a starting point to allow for comparison to the conventional configurations which are slower. It is suggested that the tracking MTEs be evaluated at several airspeeds (covering the full forward flight speed range), especially if there are airspeed based variations in the flight control system or significate variations in the plant dynamics. Special attention was applied during the development of these new tracking MTEs to allow both conventional and unconventional configurations to be capable of flying the MTEs.

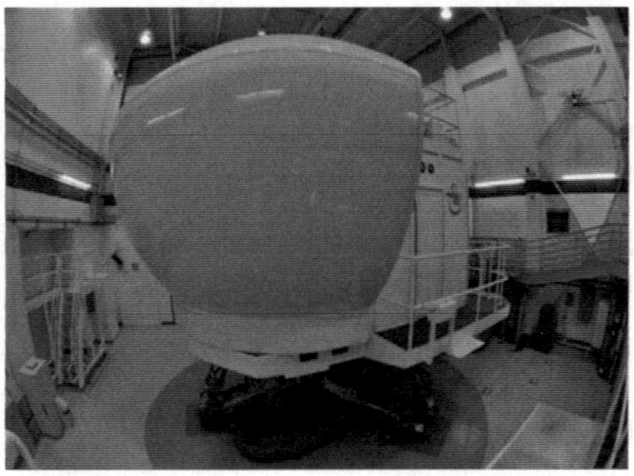

Figure 13: Sikorsky Engineering Development Simulator

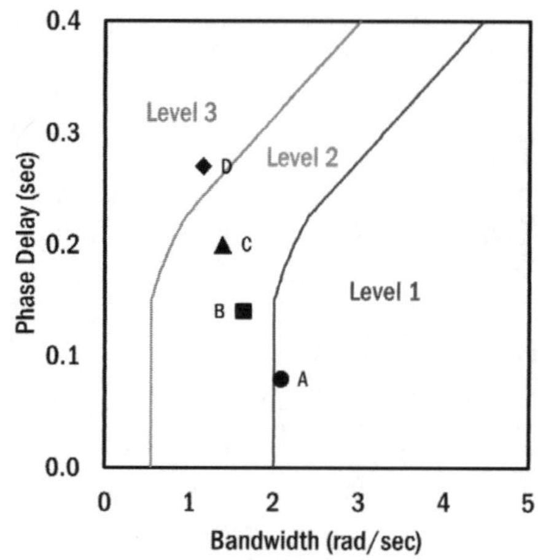

Figure 14: Small-Amplitude Pitch Attitude Changes, Forward Flight, Target Acquisition and Tracking

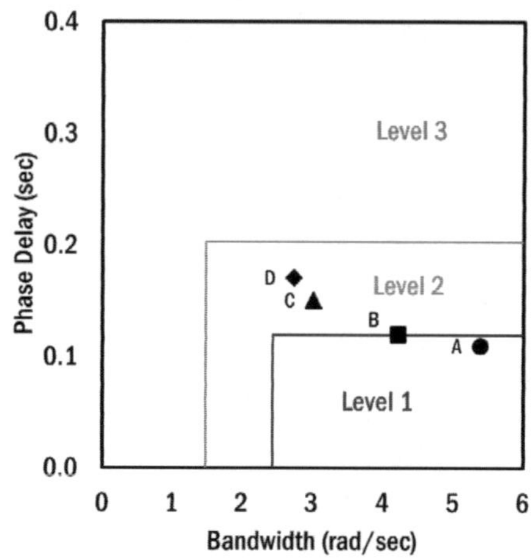

Figure 15: Small-Amplitude Roll Attitude Changes, Forward Flight, Target Acquisition and Tracking

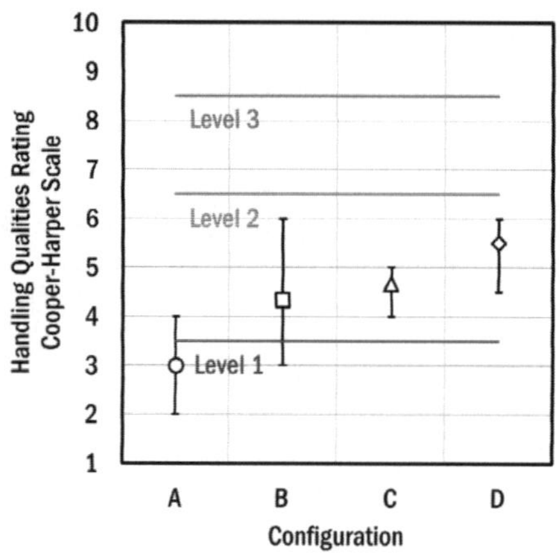

Figure 16: Depart/Abort MTE HQRs

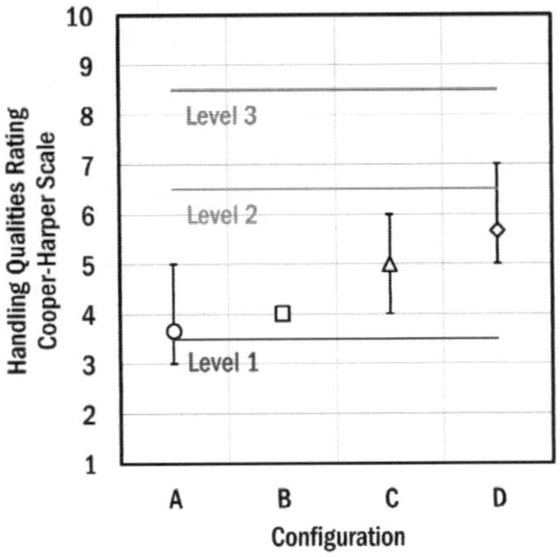

Figure 17: Slalom MTE HQRs

Sum-of-Sines Tracking MTE

The sum-of-sines tracking MTE is evaluated as a single axis task with a command signal duration of 75 seconds. The 75 seconds includes a 5 second ramp-in time, a 5 second settle time, a 60 second scoring time, and a 5 second ramp-out time that is shown in Figure 18. The pitch axis sum-of-sines command signal is shown in Figure 19. The command signal has an effective bandwidth of 0.75 rad/sec. The signal RMS is 2.5 degrees. Figure 20 depicts the command signal that is used for the roll axis. It has the same effective bandwidth of 0.75 rad/sec as pitch and a larger signal RMS of 10.0 degrees. The seven sinewaves have two different phases which are alternated from run to run to prevent the pilot from recognizing a signal pattern. The signal bandwidth and RMS are chosen based on comments and task performance from the prior pilot sessions. Values were chosen to attempt to place pilot performance within the desired criteria for exactly 50 percent of the scoring time. The signal bandwidth is similar to the command signal that is used in the first phase of Ref.9. The difference between the pitch and roll RMS values are to account for the difference in aircraft inertia between the two axes and to provide a similar level of aggressiveness for the task in each axis.

The sum-of-sines MTE requires the pilot to aggressively track the displayed signal and attempt to keep errors within the specified tolerances. The performance criteria for the SOS tracking MTE are listed below and are the same as used in Ref.15.

Desired Performance

- At least 50% of the scoring time within ±1° pitch attitude error
- At least 50% of the scoring time within ±5° roll attitude error

Adequate Performance

- At least 50% of the scoring time within ±2° pitch attitude error
- At least 50% of the scoring time within ±10° roll attitude error

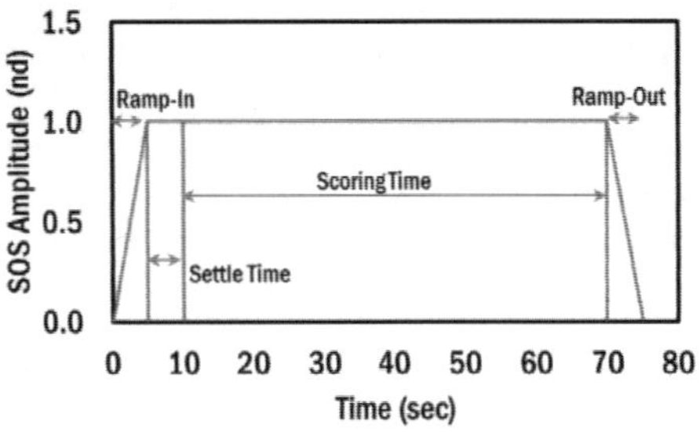

Figure 18: SOS Command Signal Structure

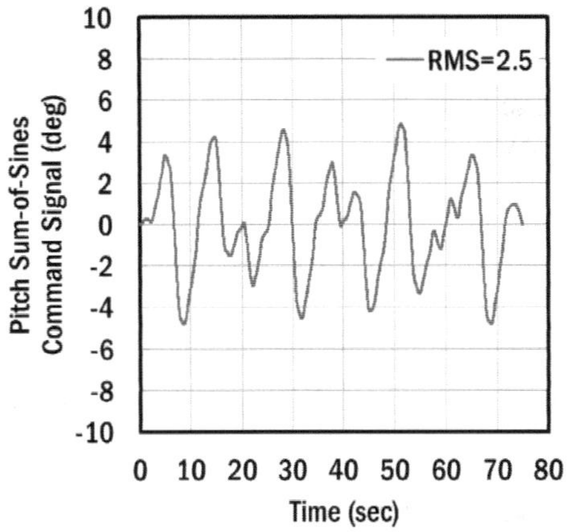

Figure 19: Pitch Axis Sum-of-Sines Command Signal, RMS 2.5 deg, Bandwidth 0.75 rad/sec

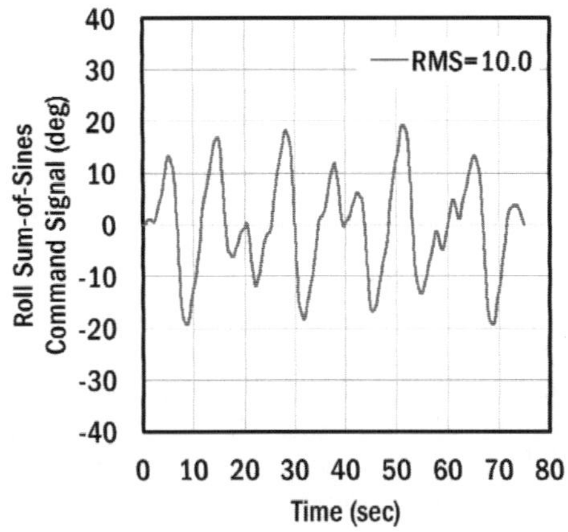

Figure 20: Roll Axis Sum-of-Sines Command Signal, RMS 10.0 deg, Bandwidth 0.75 rad/sec

Figure 21 shows the average HQRs from the evaluation pilots for each configuration for the pitch axis sum-of-sines tracking MTE compared to the Depart/Abort MTE. Figure 22 contains the correlation of the roll axis SOS tracking MTE average HQRs to the average HQRs from the Slalom MTE. The solid line in each figure indicates perfect correlation between the HQRs obtained from the two MTEs. The dotted lines depict a difference of ±1 HQR. The HQRs obtained from the pitch axis SOS MTE are well within ±1 HQR of the rating obtained from the Depart/Abort MTE. Similarly, the roll axis SOS tracking MTE shows excellent correlation (within ±1 HQR) to the HQRs obtained for the Slalom MTE.

Figure 23 and Figure 24 contain the time-on-target achieved by each pilot for the configurations evaluated. There is a clear trend of the reduction of performance from configuration A to D. The three pilots are able to achieve desired performance using configuration A for the pitch SOS tracking MTE. The time-on-target for each configuration has little variation between pilots. For the roll task, Pilot 1 and Pilot 3 are able to achieve desired performance, while Pilot 2 is slightly below desired performance for configuration A. Pilot 1 and Pilot 3 are able to achieve desired performance for configurations B and C but at an increased workload, which is reflected in the HQRs. The roll task has more variation in performance between the three pilots, but each pilot has a similar performance reduction from configuration A through D.

Figure 25 contains the pitch tracking error time history for Pilot 1 and configuration A during the scoring portion of the pitch SOS task. Figure 26 shows the roll tracking error for the roll SOS task for the same configuration and pilot. The green lines correspond to the desired performance criteria while the magenta line corresponds to adequate performance. Pilot 1 is able to track the command signal reasonably well for both tasks with only brief excursions outside of desired performance. The excursions outside of desired performance typically occur during command signal reversals after the largest attitude changes. For this example, Pilot 1 maintains desired performance for 59% of the scoring time for the pitch axis SOS tracking MTE and 80% for the roll axis task. Adequate performance or better is achieved for 90% and 95% for pitch and roll, respectively. Desired performance is achieved with minimal pilot workload and both MTEs are given a Level 1 rating by Pilot 1.

Figure 27 and Figure 28 show the open-loop frequency response for the pilot-vehicle system obtained from the pitch and roll SOS MTEs for Pilot 1 and configuration A (Ref.15). The transfer function is aircraft attitude (pitch or roll) to displayed tracking error (pitch or roll). The frequency response crossover frequency is defined as the frequency at which the pilot-vehicle system open-loop describing function magnitude crosses the 0 dB line. The crossover frequency obtained from Figure 27 is approximately 2.0 rad/sec. The value obtained from Figure 28 is approximately 1.9 rad/sec. The pilot crossover frequencies are both higher than the effective command signal bandwidth of 0.75 rad/sec. As the effective signal bandwidth is increased, the margin between the effective signal bandwidth and pilot crossover frequency will decrease for the same pilot workload. The pilot could compensate for the increase in effective signal bandwidth by increasing the workload (adding more lead) and increasing his crossover frequency. An increase in crossover frequency indicates more aggressiveness that may be driven by task, pilot or both. The crossover frequencies obtained from the pitch and roll SOS MTEs are consistent with those obtained from previous fixed-wing evaluations (Ref.9 and Ref.10).

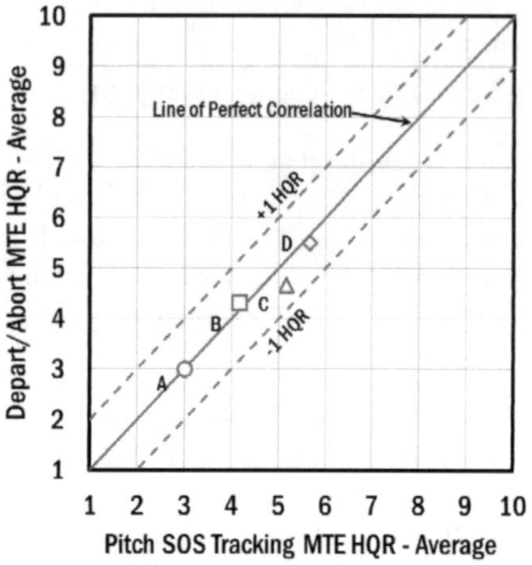

Figure 21: Correlation between Pitch SOS Tracking MTE HQR and Depart/Abort MTE HQR

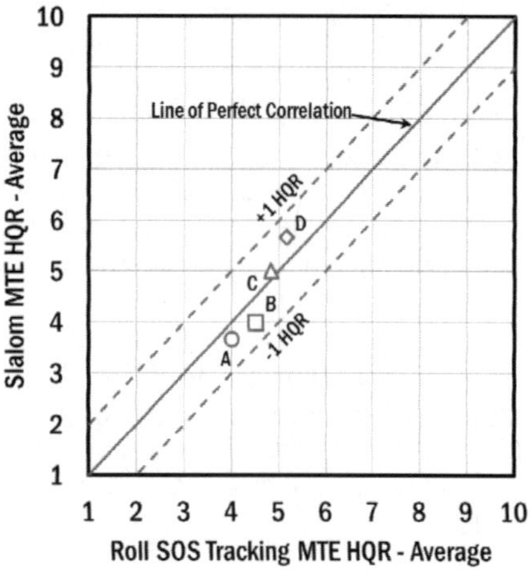

Figure 22: Correlation between Roll SOS Tracking MTE HQR and Slalom MTE HQR

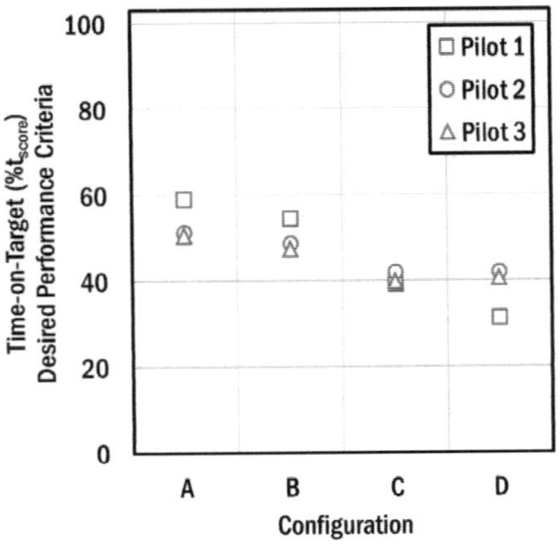

Figure 23: Pitch SOS Tracking MTE Time-on-Target

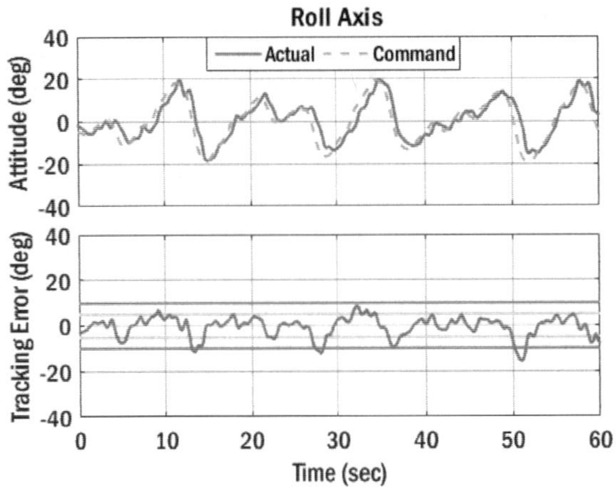

Figure 26: Roll SOS Time History, Tracking Error, Pilot 1, Configuration A

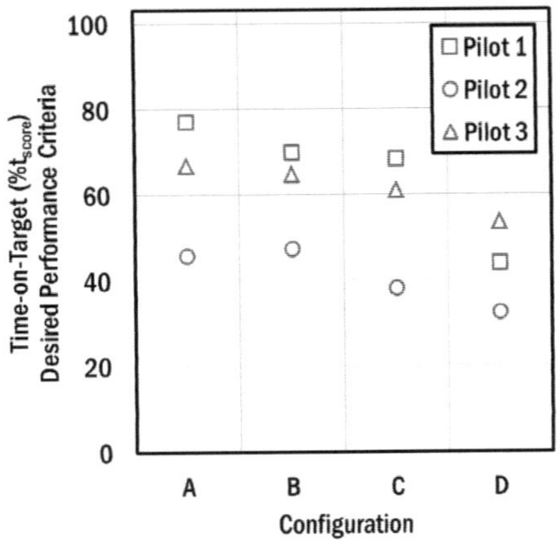

Figure 24: Roll SOS Tracking MTE Time-on-Target

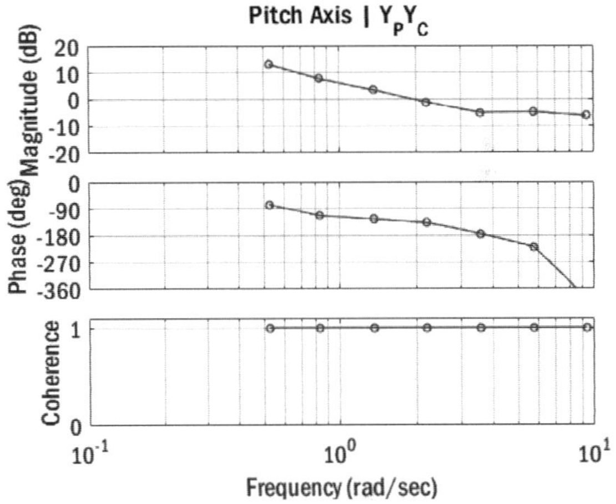

Figure 27: Pitch SOS Open-Loop Pilot-Vehicle System Frequency Response, Pilot 1, Configuration A

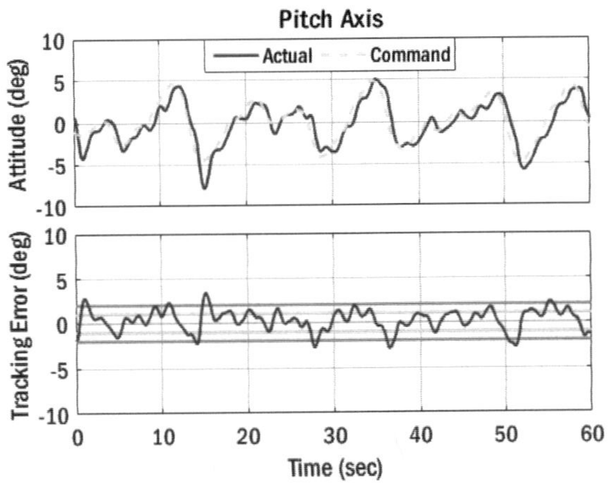

Figure 25: Pitch SOS Time History, Tracking Error, Pilot 1, Configuration A

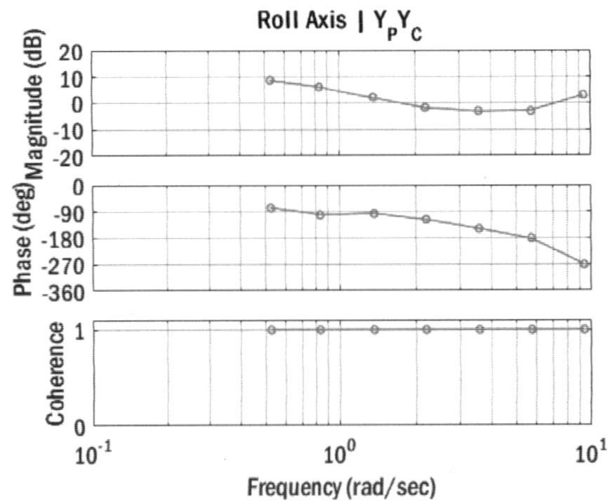

Figure 28: Roll SOS Open-Loop Pilot-Vehicle System Frequency Response, Pilot 1, Configuration A

Discrete Tracking MTE

The discrete tracking MTE is evaluated as a dual axis task. The command signal duration that is used for this evaluation is 120 seconds. The pitch and roll RMS values are 2.5 and 12.5 degrees, respectively. The signal RMS values and duration were chosen based on pilot comments during the earlier development work. An attempt was made to choose values that would place the pilot performance within the desired criteria for exactly 50 percent of the scoring time. A time-history is shown in Figure 29 and Figure 30 for the dual axis discrete tracking MTE for configuration A flown by Pilot 2. Figure 29 and Figure 30 display the command signal and tracking error for pitch and roll. The sign of the command signal is changed after each run to prevent the evaluation pilot from recognizing the signal pattern. The pitch axis command signal reaches a maximum amplitude of ±4 degrees. The roll axis command signal reaches a maximum amplitude of approximately ±20 degrees. The discrete tracking MTE used the same performance criteria as the SOS tracking MTE. Desired performance is indicated by the green lines while adequate performance is denoted by the magenta lines. Pilot 2 maintains within desired performance for 77% of the scoring time for pitch and 81% for roll. Adequate performance or better is achieved for 91% for pitch and 95% for roll. The only excursions outside of desired performance occur during the step changes in the command signal. Pilot 2 is able to quickly return inside desired performance after these brief excursions with minimal workload, thus Pilot 2 assigned a Level 1 rating for the MTE.

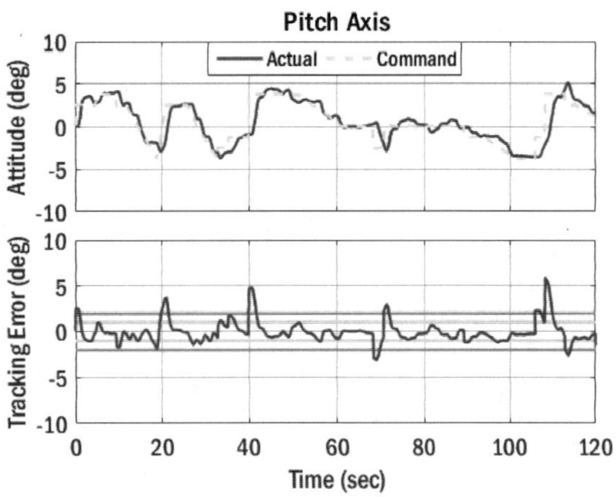

Figure 29: Discrete Tracking MTE Time History, Pitch Tracking Error, Pilot 2, Configuration A

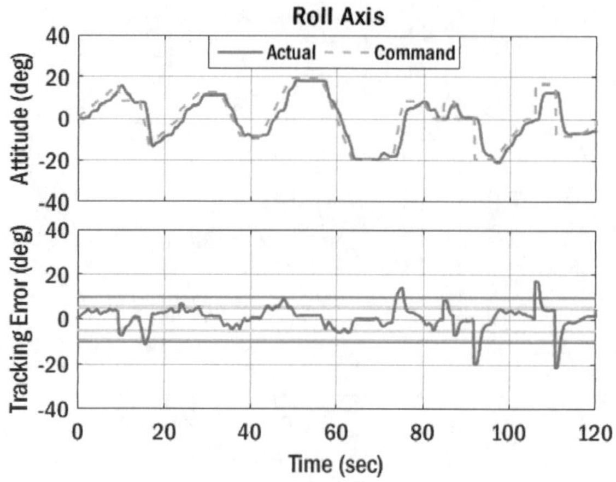

Figure 30: Discrete Tracking MTE Time History, Roll Tracking Error, Pilot 2, Configuration A

The handling qualities ratings obtained from the dual axis discrete tracking MTE are compared to the ratings from the Depart/Abort MTE and Slalom MTE in Figure 31. The solid line indicates perfect correlation between the two MTEs. The dotted line shows a deviation of ±1 HQR. The discrete tracking MTE does not show good correlation with either the Depart/Abort or Slalom MTE. The HQRs obtained from the discrete tracking MTE differ from the Depart/Abort and Slalom MTEs by more than 1 HQR. Based upon these results, the MTE, as evaluated, does not detect the known handling qualities deficiencies.

Since this discrete tracking MTE does not align with the Depart/Abort and Slalom MTEs, modifications to the task are warranted. The discrete tracking MTE may not detect the known deficiencies because it was too easy to obtain desired performance for the task and didn't require the pilot to be as tightly in the loop as necessary. The lack of pilot workload can potentially mask the handling qualities deficiencies. One potential modification could be in decrease the signal duration. This will effectively increase the signal bandwidth and increase the pilot crossover frequency. Also, increasing the RMS value of the signal will make the task more difficult. Lastly, making desired and adequate performance harder to obtain could improve the MTE's ability to identify handling qualities deficiencies since it will require the pilot to be more in-the-loop. Each of these knobs should be investigated in future work before the task is eliminated from consideration.

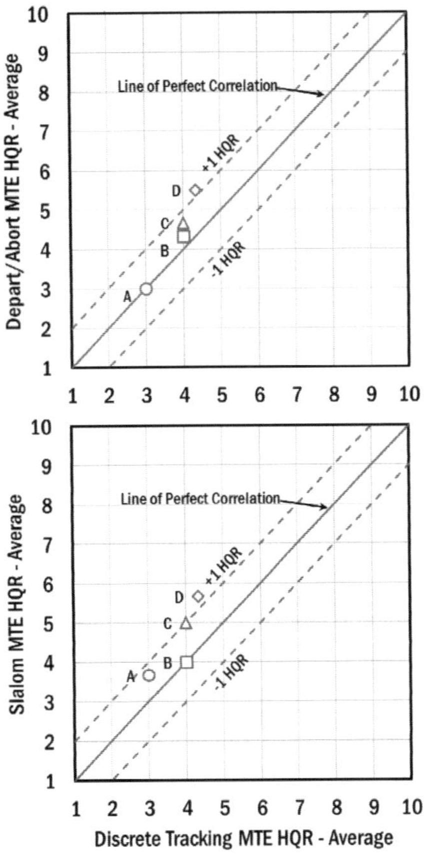

Figure 31: Correlation between Discrete Tracking MTE HQR, Depart/Abort MTE HQR, and Slalom MTE HQR

Figure 32 and Figure 33 show the time-on-target for pitch and roll for each of the configurations. The pilots are able to achieve desired performance in both axes. In addition, the three pilots obtained similar performance during the evaluation of each configuration. The time-on-target degrades slightly between configurations. The slight reduction in performance from configuration A to D suggests that the MTE successfully reveals some of the handling qualities deficiencies, but the degradation is not as much as shown by the SOS tracking MTEs. The performance reduction may be magnified if the pilot is required to be more tightly in the loop. This is supported by the crossover frequency of the pilot-vehicle system obtained for pitch and roll for the dual axis discrete tracking MTE. The crossover frequency for pitch is approximately 0.7 rad/sec and 1.0 rad/sec for roll. The crossover frequency from the discrete MTE is significantly lower than values obtained from the sum-of-sines tracking MTEs.

The discrete tracking MTE could be made more challenging by decreasing the command signal duration, which effectively increases the command signal bandwidth for a constant RMS value. It is recommended to re-evaluate the discrete tracking MTE with a shorter duration before removing the MTE from consideration.

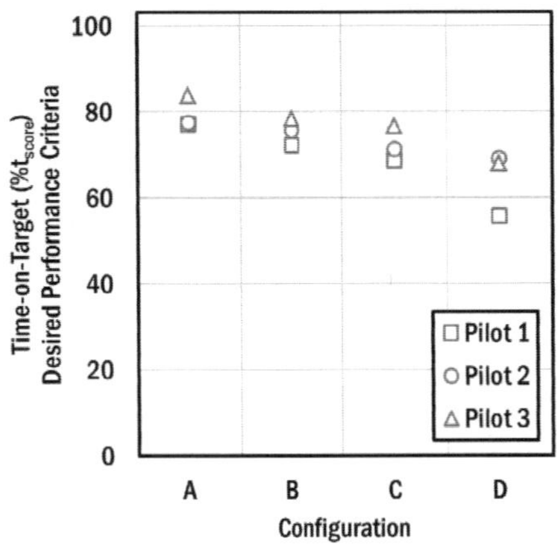

Figure 32: Discrete Tracking MTE Time-on-Target, Pitch

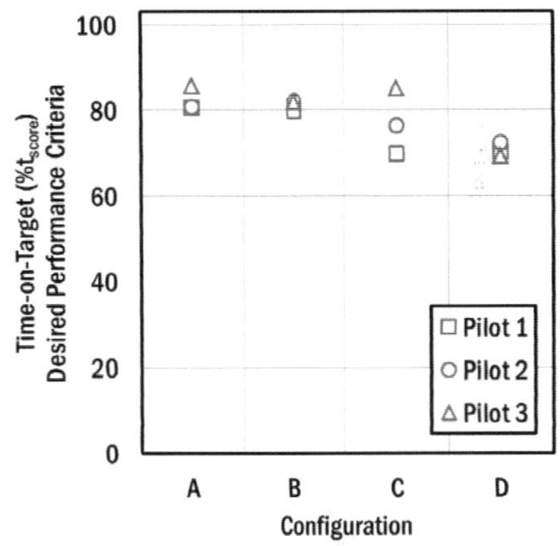

Figure 33: Discrete Tracking MTE Time-on-Target, Roll

CONCLUSIONS

As the requirements for FVL push rotorcraft technology development, a need arises in ADS-33 to ensure these new designs demonstrate desired handling qualities throughout their operational flight envelope. The present work shows how two tasks, flown for decades by the fixed-wing community, can be modified for rotorcraft applications. These two tasks, sum-of-sines and discrete tracking, fit into the mission-oriented approach already established by ADS-33E-PRF and fulfill the precision, aggressive MTE category.

From this study, the following conclusions have been drawn:

- Sum-of-sines tracking task performance and workload can be modified by command signal bandwidth and RMS value.

- Discrete tracking task performance and workload can be modified by command signal duration and RMS value.
- The HQRs obtained from the sum-of-sines tracking MTE are within ±1 HQR of the Depart/Abort and Slalom MTEs for each configuration evaluated.
- The sum-of-sines tracking MTE evoked higher levels of aggressiveness in the task than the discrete tracking MTE.
- The HQRs obtained from the discrete tracking MTE, using a 120 second signal duration, did not correlate well (more than ±1 HQR difference) with either the Depart/Abort or Slalom MTEs.
- The discrete tracking MTE correlation could potentially be improved by increasing the level of aggressiveness of the task by reducing the signal duration, increasing the command signal magnitude or making desired performance harder to obtain.

ACKNOWLEDGEMENTS

This effort is partially sponsored by the U.S. Government under Other Transaction number W15QKN-10-9-0003 between Vertical Lift Consortium, Inc. and the Government. The authors are grateful to the U. S. Army's National Rotorcraft Technology Center (NRTC) program and Sikorsky, a Lockheed Martin Company for co-funding this research under the Base VLC Project Agreement No. 2015-327. The US Government is authorized to reproduce and distribute reprints for Government purposes notwithstanding any copyright notation thereon The views and conclusions contained herein are those of the authors and should not be interpreted as necessarily representing the official policies or endorsements, either expressed or implied, of the U.S. Government.

The authors would also like to thank Mr. Chris Flaws, Mr. Brian Wach, Mr. Kerr Olivier, and Mr. Evan Dobbins of Sikorsky Simulation Group for their support on visual development and simulation integration. In addition, the authors would like to thank LTC Carl Ott and Mr. John Wood of the U.S. Army's Aviation Development Directorate – AATD for their support in the piloted evaluations.

REFERENCES

[1]"Aeronautical Design Standard, Performance Specification: Handling Qualities Requirements for Military Rotorcraft," ADS-33E-PRF, U.S. Army Aviation and Missile Command, March 2000.

[2]Mitchell, D. G., R. H. Hoh, B. L. Aponso, and D. H. Klyde, *Proposed Incorporation of Mission-Oriented Flying Qualities into MIL-STD-1797A*, WL-TR-94-3162 October 1994.

[3]Cooper, George E., and Robert P. Harper, Jr., *The Use of Pilot Rating in the Evaluation of Aircraft Handling Qualities*, NASA TN D-5153, April 1969.

[4]Klyde, D. H., B. L. Aponso, and D. G. Mitchell, *Handling Qualities Demonstration Maneuvers for Fixed-Wing Aircraft, Volume I: Maneuver Development Process*, WL-TR-97-3099, October 1997.

[5]"Flying Qualities of Piloted Aircraft," MIL-STD-1797B Notice 1, Department of Defense, April 2012.

[6]Blanken, C. L., Tischler, M. B., Lusardi, J. A., Ivler, C. M., "Aeronautical Design Standard – 33 (ADS-33)...Past, Present, and Future," AHS Rotorcraft Handling Qualities Specialists' Meeting, Huntsville, AL, April 2014.

[7]King, R., and Wiggins, L., "Draft Changes to ADS-33E-PRF," AHS Rotorcraft Handling Qualities Specialists' Meeting, Huntsville, AL, April 2014.

[8]McRuer, D. T., and E. S. Krendel, *Mathematical Models of Human Pilot Behavior*, AGARDograph No. 188, November 1973.

[9]Bailey, R. E., and Knotts, L. H., *Interaction of Feel System and Flight Control System Dynamics on Lateral Flying Qualities*, NASA CR 179445, 1990.

[10]Bailey, R. E., Knotts, L. H., and Levison, W. H., "An Investigation of Time Delay During In-Flight and Ground-Based Simulation," Calspan Report No. 7205-16, Vols. I and II, August, 1987.

[11]Neal, T. Peter, and Rogers E. Smith, *An In-Flight Investigation to Develop Control System Design Criteria for Fighter Airplanes*, AFFDL-TR-70-74, Vol. I, December, 1970.

[12]Monagan, Stephen J., Rogers E. Smith, and Randall E. Bailey, *Lateral Flying Qualities of Highly Augmented Fighter Aircraft*, AFWAL-TR-81-3171, June 1982.

[13]Sahasrabudhe, V., Faynberg, A., Kubik, S., Tonello, O., Pritchard, J., and Gubbels, A., "CH-53K Control Laws: Risk Reduction Flight Testing," *Proceedings of the American Helicopter Society 67th Annual Forum*, Virginia Beach, VA, May 2011.

[14]Kothmann, B., and Armbrust, J., "RAH-66 Comanche Core AFCS Control Law Development: DEMVAL to EMD," *Proceedings of the American Helicopter Society 58th Annual Forum*, Montreal, Canada, June 2002.

[15]Klyde, D. H., Schulze, P. C., and Thompson, P. M., "Exposing Unique Pilot Behavior from Flight Test," *AIAA Atmospheric Flight Mechanics Conference*, Kissimmee, FL, January 2015.

A Flight Path Generation Algorithm for the Development and Preliminary Assessment of Mission Task Elements

Adam T. Thorsen
Graduate Research Assistant
The Pennsylvania State University
University Park, PA 16802

Joseph F. Horn
Professor
The Pennsylvania State University
University Park, PA 16802

ABSTRACT

A flight path generation (FPG) algorithm that calculates control inceptor commands required for complex maneuvering flight was developed. This tool was used in previous work to simulate non-piloted aerobatic maneuvers of a conventional and compound H-60. The flight path tool models a generalized flight path with a series of straight and turning segments. A mode that generates minimum time flight paths using load factor and roll rate constraints is the focus. This study seeks to extend the application of this flight path tool to the development and assessment of mission task elements for next generation rotorcraft. In particular, the rotorcraft model examined is a notional compound H-60 featuring a wing and pusher propeller. A basic flight control system was developed using Nonlinear Dynamic Inversion (NLDI) control architecture. Two flight demonstration maneuvers are examined in piloted simulation: a break turn and a descent turn. The break turn is a high speed turning maneuver where the objective is to turn through 90 deg. as quickly as possible. The flight path tool can provide minimum time criteria given an airspeed range, maximum vertical load factor, and maximum roll rate. The descent turn is similar to a wind-up turn and requires the pilot to perform a spiraling dive through a target change in horizontal flight path angle. Although specific performance metrics were not explored for this maneuver, the FPG tool was used to regenerate the piloted flight path and thus demonstrate the ability of the FPG to accurately replicate generalized piloted maneuvers.

NOMENCLATURE

ACAH	attitude command attitude hold
ADS	aeronautical design standard
CD_p	primary control derivative matrix
CD_r	redundant control derivative matrix
EOMs	equations of motion
FM	forces and moments vector
FM_0	trimmed forces and moments vector
FPG	flight path generation
GENHEL	general helicopter
JMR-FVL	joint multi-role future vertical lift
MTE	mission task element
NLDI	nonlinear dynamic inversion
PI	proportional-integral
PID	proportional-integral-derivative
RCAH	rate command attitude hold
RCTC	rate command with turn coordination
SD	stability derivative matrix
VTDP	vectored thrust ducted propeller
a_y	lateral acceleration
g	acceleration due to gravity
h	altitude
N_z	vertical load factor
p	roll rate
r	yaw rate
R_c	turn radius in segments 23 and 56
R_e	effective turn radius
V	airspeed
X_e, Y_e, Z_e	earth-fixed position coordinates
γ	vertical flight path angle
δ	control input
θ	pitch attitude
χ	horizontal flight path angle
act	actual
coll	main rotor collective
cmd	command
e	error
flap	wing flaperon
lat	lateral cyclic
long	longitudinal cyclic
max	maximum
ped	pedals
prop	propeller
targ	targ
thr	throttle

Presented at the American Helicopter Society Rotorcraft Handling Qualities Technical Meeting, Huntsville, AL, Feb. 22 - 23, 2017.

INTRODUCTION

Conventional rotorcraft task a pilot with the challenge of managing four primary controls (main rotor collective, longitudinal cyclic, lateral cyclic, and tail rotor collective) while monitoring a variety of flight envelope limitations, such as power, structural limits, and vibrations. With compound rotorcraft, additional control effectors (e.g. wing flaperons for auxiliary lift and propeller pitch for auxiliary thrust) are introduced, which can be manually controlled by the pilot or automatically allocated during flight.

Compound rotorcraft have high speed and maneuvering capabilities with the potential to significantly surpass those of conventional rotorcraft. Compound rotorcraft can overcome several conventional rotorcraft limitations at high speeds with the tradeoff of increased complexity, weight, and degraded low speed and hover performance. Rotorcraft compounded with pusher propellers require a monitoring of main rotor collective and propeller pitch allocation based on the thrust required in a maneuver. The addition of a wing requires an understanding of the lift sharing between the main rotor and wing. Wing flaperons affect the workload with regard to the force and moment generations required from all four of the primary controls of a conventional rotorcraft. Flaperons deflected in a differential role can influence lateral cyclic through rolling moment contributions by overcoming some of the wing's natural roll damping.

With a comprehensive understanding of the influence of auxiliary forces and moments introduced by redundant controls, simulations of trim and basic maneuvering flight are necessary to assess controller performance, vehicle performance, and handling qualities. Investigations of controller performance in complex maneuvering flight (e.g. aerobatics associated with combat maneuvers) are important in determining the appropriateness and robustness of the control architecture.

A generalized method of producing command profiles characteristic of any maneuver is desirable for conducting (non-piloted) dynamic simulations in an efficient manner. With generalized maneuver simulations, an assessment of the maneuvering potential of a compound rotorcraft can be made. This is essential for the development of new qualitative and quantitative criteria, along with mission task elements (MTEs) for next generation rotorcraft in future ADS-33 releases. MTEs need to capture handling quality deficiencies of these next generation rotorcraft.

Understanding the effectiveness of auxiliary lift and propulsion as well as redundant controls on compound rotorcraft is important. Developing a practical flight control system accommodating these controls is critical for next generation rotorcraft, such as those featured in the JMR-FVL Program; this program has several competing participants and rotorcraft configurations with varying degrees of compounding and control redundancy. Two configurations have been downselected (Defiant, V-280). Sikorsky-Boeing's Defiant has a pusher propeller while Karem's Optimum Speed Tiltrotor has variable RPM capabilities. AVX's entrant features a coaxial main rotor system and two ducted pusher propellers mounted on a stub wing along with a canard, which introduces aspects of both lift and thrust compounding. Bell's V-280 Valor is a tiltrotor concept aimed at achieving a cruise speed of 280 knots and also features a V-tail structure with ruddervators.

DARPA's X-VTOL program, which calls for future rotorcraft to have hovering efficiencies greater than 75% and high speed capabilities of 300 - 400 knots, also features compound rotorcraft configurations with redundant controls. Boeing's Phantom Swift has wing tip mounted ducted pusher propellers and lift fans integrated into the fuselage, which introduces control redundancy in the form of lift and thrust compounding. Aurora's LightningStrike X-Plane, which was awarded the DARPA contract, features distributed ducted fan propulsion with a tilt-wing and canard. Other participants included Karem's tiltrotor concept and Sikorsky's Rotor Blown Wing concept.

Previous work has been conducted involving compound rotorcraft with control redundancy and the effect on rotorcraft power, vibrations, and flight controls. Stabilators were studied for potential reductions in hub vibration levels in high speed forward flight [1, 2]. The performance benefits of variable rotor speed has also been explored [3 - 7]. A maneuver optimization and controller study for a compound rotorcraft was performed using a UH-60A with a wing and vectored thrust ducted propeller (VTDP). An optimal schedule was incorporated into a Fly-by-Wire controller, and accelerations and pull-ups were simulated [8, 9].

A flight control system was designed to conduct in-flight optimization of redundant control effectors using Adaptive Performance Optimization (APO) [10]. APO uses low frequency forced responses to measure performance parameters and was originally developed for fixed-wing transport aircraft [11 - 14]. This work was improved and extended to multi-variable trim optimization on a compound rotorcraft by performing a two-dimensional search through redundant controls using a Fly-to-Optimal Concept [15, 16].

Optimizing control in quasi-steady maneuvering flight was investigated in order to minimize power in high speed accelerations, pull-ups, and generalized maneuvers using redundant controls on a compound rotorcraft and a linear dynamic inversion controller [17, 18]. This work was expanded to the use of redundant controls in enhancing transient response and handling qualities of a compound rotorcraft [18, 19] and was inspired by work done by Pavel and Padfield, which sought to extend existing ADS-33 metrics and enhance the effectiveness of multi-disciplinary design [20]. Differential wing flaps demonstrated the most potential for elevating roll attitude quickness values in transient lateral maneuvers while symmetric wing flaps allowed the aircraft to achieve high levels of agility quickness.

Past work containing applications of a nonlinear dynamic inversion controller (which is the control architecture used in this study) in simulation testing of advanced response types for ship-based rotorcraft and investigations of bandwidth and disturbance rejection properties was performed [21 - 24].

After investigations of basic maneuvers and the initial development of an improved and more robust controller (using the NLDI control architecture), attention was shifted to simulating more complex maneuvers. Aerobatic maneuvers, which can be MTEs, are particularly cumbersome and difficult to simulate. A generalized method of producing command profiles characteristic of any maneuver is desirable for conducting simulations in an efficient manner. Work involving flight path generation has been approached with various methodologies, depending on application. A trajectory parameterization method for calculating emergency flight paths in real-time was developed via modified Dubins parameterization [25, 26]. Dubins curve was modified to allow for (piecewise constant) acceleration along the path and a finite rate of change in turn rate. A number of closed-form solutions were developed for path-defining variables, and the path was calculated to minimize a performance objective (e.g. desired touchdown location) subject to specified constraints (e.g. aircraft states). Additional flight path generation work has focused on more representative functions for path-defining variables and less on real-time path planning [27, 28]. These methodologies use different order polynomials to represent a selected set of variables such as airspeed, altitude, turn rate, and/or load factor. Again, performance objectives such as desired terminal position coordinates can be targeted in generating the flight path. This type of inverse simulation relies on numerical techniques to calculate control allocations to satisfy a particular trajectory. These flight paths were generated to model ADS-33 maneuvers in order to investigate the maneuverability of two helicopter configurations.

Recent work focused on very complex and demanding maneuvers using an NLDI control architecture for a compound rotorcraft [29 - 31]. A flight path generation algorithm was developed to produce control inceptor commands in order to simulate aerobatic maneuvers. The NLDI controller was also adapted for a baseline H-60 to allow the aerobatic maneuvers to be simulated and compared for both a compound and conventional rotorcraft.

Work continued with the NLDI controller in aerobatic maneuvering flight at high advance ratios. The combat break turn (CBT) was one of the maneuvers examined and is shown in Fig. 1; this maneuver is a time-sensitive 90 deg. turn [32]. The CBT is an important maneuver to examine, especially for future compound MTEs, since it involves a common requirement which is present in current MTEs — namely assessing the aircraft's maneuvering capability via time requirements. Depending on the situation, the pilot may choose to perform the CBT with or without airspeed bleed-off.

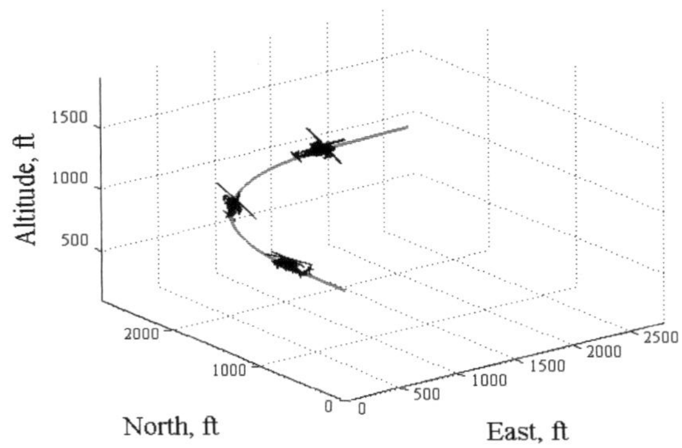

Figure 1. Break Turn.

The other MTE introduced in this work is a combat descent turn (CDT) and is shown in Fig. 2. The combat descent turn (CDT) is a maneuver in which the aircraft performs a spiraling dive; CDT illustrates a primarily defensive combat maneuver and is inspired by fixed-wing defensive spiral and sliceback maneuvers. A fixed-wing sliceback maneuver is characterized as a 180 deg. descending turn in which 135 deg. of bank is achieved. This maneuver is also similar to a wind-up turn, which is a descending maneuver of increasing bank angle and decreasing turn radius [33, 34].

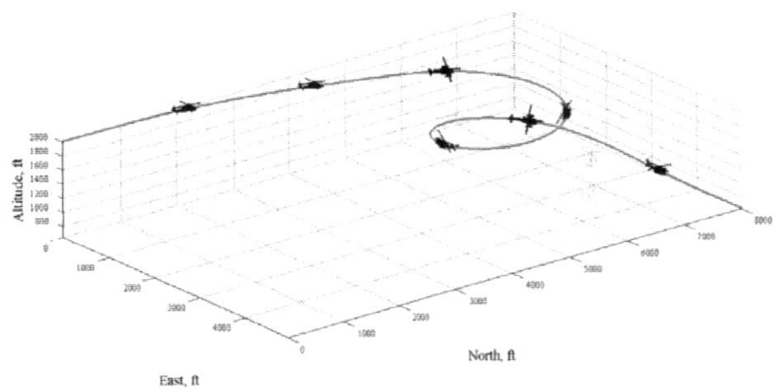

Figure 2. Descent Turn.

This study seeks to present a generalized and efficient method to aid in the development and assessment of MTEs and introduce maneuvers useful for capturing handling qualities on multiple control axes. The two proposed compound MTEs that will be presented are a break turn and a descent turn.

This paper is organized into the following sections: Simulation Environment, Flight Path Generation, Flight Control Design, Break Turn Simulations, Descent Turn Simulations, Conclusions, and Future Work.

SIMULATION ENVIRONMENT

The core of the simulation model is a nonlinear blade element mathematical model of the UH-60A Black Hawk, which was developed from the U.S. Army General Helicopter (GENHEL) Flight Dynamics Simulation. GENHEL is well-established and has been validated with flight test data [35]. The version of GENHEL used in this work was developed at Pennsylvania State University (GENHEL-PSU) for research purposes in the areas of flight dynamics and control [36]. GENHEL models the six rigid body degrees of freedom, along with four rotor flapping and four rotor lagging degrees of freedom. Dynamic states due to main rotor inflow are modeled with a 33 state Peters-He inflow model.

Simulations of a notional compound rotorcraft are used. The aircraft is based on an H-60 airframe and includes a wing and pusher propeller model that were developed in GENHEL-PSU. Lookup tables with finite wing corrections are used for the wing's two-dimensional lift, drag, and pitching moment coefficients based off of an NACA 63-412 airfoil. The compound H-60 model is illustrated in Fig. 3. Table 1 presents a brief summary of the specifications of the compound H-60.

Figure 3. Compound Rotorcraft Configuration.

Table 1. Compound Rotorcraft Specifications.

Vehicle General	
Gross Weight	20,000 lbs.
Available Power	4,500 HP
Main Rotor	
Radius	26.83 ft.
Chord (0.75R)	1.73 ft.
RPM (100% N_r)	258 RPM
Number of Blades	4
Propeller	
Diameter	8 ft.
RPM (100% N_r)	2160
Number of Blades	7
Wing	
Effective Span	45 ft.
Effective Wing Area	226 ft^2
Mean Aerodynamic Chord	5.02 ft.

Propeller Model

Blade element momentum theory was combined with momentum theory and Goldstein's classical vortex theory in order to create a flight dynamic model of the pusher propeller [31, 37].

The propeller model is for a constant speed and variable pitch propeller. Along with all the basic inputs for a propeller model (number of blade elements, diameter, number of blades, RPM, spinner diameter, radial chord and geometric twist variations, thickness/chord ratio, etc.) there are options to set the angle of attack and sideslip angle of the propeller disk.

Airfoil data for a Clark-Y is built into the code using local Reynolds number computations to drive lift slope, zero lift drag coefficient, and stall angle of attack variations. Aspects of basic vortex theory are combined with blade element momentum theory. Vortex theory is integrated as corrections to the sectional angle of attack for flow curvature and blade thickness. Quasi-steady lift and drag coefficients are used for high local angles of attack.

Optimization methods are applied to drive two components of inflow (perpendicular and tangential with respect to the propeller plane) at each blade element and azimuth angle to convergence. These iterated components of inflow model the helical trailing-edge vortices shed downstream of the propeller plane of rotation.

The optimization methods are adaptive and were developed to improve convergence characteristics of the theory for reverse thrust cases. Modified Newton's Method is used with an approximate line search featuring a variable step size adhering to Wolfe's conditions of optimality. Additional logic was developed to align restart criteria with modified initial inflow guesses and search directions.

Flight Simulator

The rotorcraft flight simulator at the Penn State Vertical Lift Research Center of Excellence (VLRCOE) uses the cockpit of a Bell XV-15 tiltrotor prototype with custom instrumentation and cab displays. Force-feel characteristics are provided through the control inceptors via a 300 lb. capable control loading system, and X-Plane software is used to generate the flight simulation scenery. Three ceiling mounted projectors are used to provide a 170 deg. field of view onto a cylindrical screen with a 15 ft. diameter. The cockpit of the flight simulator has a gaming throttle on the center console; this throttle is based on the A-10 throttle inceptor. This fore/aft throttle was used as the pilot's left hand inceptor along with the collective stick inceptor for the piloted simulations in this study.

FLIGHT PATH GENERATION

Flight Path Discretization

A flight path generation (FPG) algorithm was developed as a software tool in the MATLAB environment [27 - 31]. Turning flight paths generated using load factor and roll rate constraints, and straight segments governed by distance-based objective functions are the focus of this study. The flight path generation is kinematics-based and uses the flight path structure shown in Fig. 4 for an earth-fixed frame of reference. The flight path is discretized into eight different segments. The flight path from points 2 through 4 and 4 through 7 are the turning segments in which altitude (*h*) and airspeed (*V*) can be specified. The north-east plane exit coordinates of the turn trajectory are computed using the horizontal flight path angles covered in each turn segment along with each turn's effective turn radius as shown in Eqs. 1 where *sgn* is the signum function.

$$X_{e,\text{exit}} = X_{e,\text{entry}} + R_e \sin(\chi_{\text{turn}})$$
$$Y_{e,\text{exit}} = Y_{e,\text{entry}} + \text{sgn}(\chi_{\text{turn}})[R_e(1 - \cos(\chi_{\text{turn}}))] \quad (1)$$

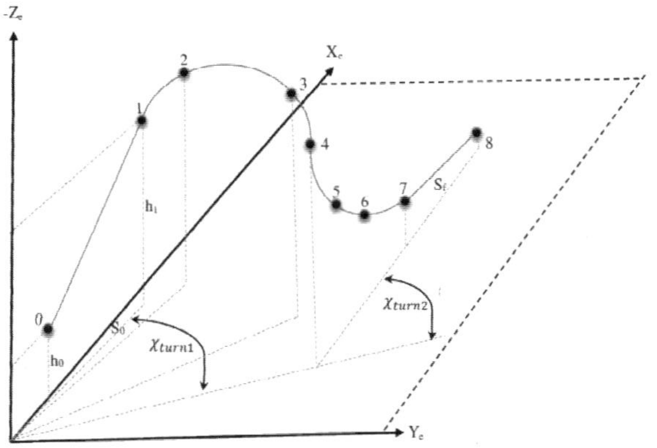

Figure 4. Generalized Flight Path Discretization.

A scalar factor is applied to the turning segments from points 1 to 2, 3 to 4, 4 to 5, and 6 to 7. This factor controls the amount of time spent by the aircraft in rolling to the initial bank angle and rolling out of the final bank angle in the turn segments. For symmetric steady turns, a desired maximum time rate of change of the bank angle ($\dot{\phi}_{\max}$) can be achieved by solving the system (via MATLAB's **`fsolve`**) shown in Eqs. 2 for the time at maximum roll rate and the horizontal flight path angle traversed in the roll-in or roll-out segment.

$$\dddot{\chi}(g^2 + V^2\dot{\chi}^2) - 2V^2\dot{\chi}\ddot{\chi}^2 = 0$$
$$\frac{V\ddot{\chi}}{g + \frac{(V\dot{\chi})^2}{g}} - \dot{\phi}_{\max} = 0 \quad (2)$$

A total turn angle (χ_{turn}) is specified, which is used in Eq. 1. This turn angle for each turn segment is divided into three parts, including the aforementioned roll-in and roll-out segments (e.g. segments 12, 23, and 34). The scalar factor (ξ_T) controls the roll rate in the roll-in (e.g. segment 12) and roll-out (e.g. segment 34) segments, and an asymmetric factor (ς) allows for a different turn angle to be traversed in the roll-in and roll-out segments. The horizontal flight path angles traversed (for the first turn segment) are discretized for a turn segment given χ_{turn}, ξ_T, and ς, where the greatest integer function is used to determine the number of turns (n_T); this is shown in Eqs. 3 - 5. Note that $\xi_T \in (0, 0.5)$ and $\varsigma \in (-1, 1)$. Each of these horizontal flight path angles is used to solve the systems of equations that govern each of the eight total flight path segments (i.e. these parameters are used in solving the turn rate segment variations which are given by Eqs. 8 or 10).

$$n_T = \left\lfloor \frac{|\chi_{\text{turn}}|}{2\pi} \right\rfloor \quad (3)$$

$$\begin{cases} \chi_{\text{turn}}' = \text{sgn}(\chi_{\text{turn}})[2\pi(n_T + 1) - |\chi_{\text{turn}}|] \\ \chi_{12} = \xi_T(\varsigma + 1)[\chi_{\text{turn}}' + 2\pi(n_T + 1)] \\ \chi_{23} = (1 - 2\xi_T)[\chi_{\text{turn}}' + 2\pi(n_T + 1)] \\ \chi_{34} = \xi_T(1 - \varsigma)[\chi_{\text{turn}}' + 2\pi(n_T + 1)] \end{cases}, \quad \chi_{\text{turn}} < 0 \quad (4)$$

$$\begin{cases} \chi_{\text{turn}}' = \chi_{\text{turn}} \\ \chi_{12} = \xi_T(\varsigma + 1)\chi_{\text{turn}}' \\ \chi_{23} = (1 - 2\xi_T)\chi_{\text{turn}}' \\ \chi_{34} = \xi_T(1 - \varsigma)\chi_{\text{turn}}' \end{cases}, \quad \chi_{\text{turn}} \geq 0 \quad (5)$$

Mathematical Functions for Path-Defining Variables

The altitude, airspeed, and turn rate variations in each segment are determined in part using MATLAB's **`fsolve`**, which uses trust-region-dogleg optimization to solve for unknown coefficients in systems of nonlinear equations. These unknown coefficients (a_v ... d_v, a_h ... d_h, a_k ... d_k) are unique for each flight path segment and are paired with the airspeed, altitude, and turn rate time histories. Each of these time histories are modeled with polynomials of different orders (depending on the flight path segment) as shown in Eqs. 6 - 8. The unknown coefficients are determined by **`fsolve`** using inputs and information from preceding flight segments as boundary conditions.

$$V(t) = a_v t^3 + b_v t^2 + c_v t + d_v \quad (6)$$

$$h(t) = a_h t^3 + b_h t^2 + c_h t + d_h \quad (7)$$

$$\dot{\chi}(t) = a_k t^3 + b_k t^2 + c_k t + d_k \quad (8)$$

The order of the polynomial representing airspeed, altitude, or turn rate in a specific segment was determined based on known variations of altitude, airspeed, and turn rate from piloted simulations performed in previous work for generalized maneuvers [18, 31]. Previous versions of the FPG coupled turn rate and airspeed in segments 23 and 56 [29 - 31]. An option to decouple turn rate and airspeed was developed in the FPG tool for a more accurate (namely, less constrained) representation of turn rate in bleed-off maneuvers (e.g. break turns). Load factor constraints are directly incorporated into Eq. 8 as boundary conditions (at points 2, 3, 5, and 6), using the relationship in Eq. 9.

When airspeed and turn rate are coupled, the turn rate is governed by Eqs. 10, and load factor constraints are implemented as penalty functions into a time-based objective function for which a feasible region of turn radii in segments 23 and 56 is searched.

$$\dot{\chi} = \frac{g\sqrt{{N_z}^2 - 1}}{V} \tag{9}$$

$$\begin{gathered}\dot{\chi}_{12}(t), \dot{\chi}_{34}(t), \dot{\chi}_{45}(t), \dot{\chi}_{67}(t) = a_k t^3 + b_k t^2 + c_k t + d_k \\ \dot{\chi}_{23}(t) = V_{23}(t)/R_{c1}\,, \dot{\chi}_{56}(t) = V_{56}(t)/R_{c2}\end{gathered} \tag{10}$$

Additional important equations used in the algorithm are outlined in Eqs. 11 - 13.

$$\begin{gathered}\dot{X}_e = Vcos(\gamma)cos(\chi) \\ \dot{Y}_e = Vcos(\gamma)sin(\chi) \\ \dot{Z}_e = -\dot{h}\end{gathered} \tag{11}$$

$$\gamma = -\sin^{-1}\left(\frac{\dot{Z}_e}{V}\right), \dot{\gamma} = \frac{\dot{V}\dot{Z}_e - \ddot{Z}_e V}{V^2 \cos(\gamma)} \tag{12}$$

$$\phi = \sin^{-1}\left(\left(\frac{V}{g}\right)\dot{\chi}\cos(\gamma)\Big/\sqrt{1 + \left(\frac{V}{g}\dot{\chi}\right)^2}\right) \tag{13}$$

Optimization Methods

The vertical load factor constraints are inputs along with maximum roll rate that dictate the effective turn radius (and hence the end coordinates governed by Eq. 1) in turning segments for a time-based objective function.

The times spent in each segment are solved based on inputs and optimization methods. The time in each of the straight segments (covering a specified ground distance, *S*) is determined using a golden section search.

Golden section is a one-dimensional bracketing method [38]. A parameter known as the golden ratio (τ) is used to adjust search points within the current interval. With absolute upper (b) and lower (a) bounds defining the feasible region (i.e. time for this study), the initial search interval [x_1, x_2] is computed using Eq. 14.

$$\begin{gathered}x_1 = a + (1 - \tau)(b - a) \\ x_2 = a + \tau(b - a) \\ \tau = \frac{\sqrt{5} - 1}{2}\end{gathered} \tag{14}$$

The objective function, *f*, (i.e. the error between the target and actual ground distance for this study) is computed at each interval bound for a particular iteration. The algorithm reaches a stopping criteria when the interval becomes sufficiently small based on some prescribed tolerance. The interval [x_1, x_2] is continually adjusted based on the logic given in Eqs. 15 and 16 (for a minimization problem) depending on the value of the objective function as the iteration progresses.

$$\begin{cases} b = x_2 \\ x_2 = x_1 \\ x_1 = a + (1 - \tau)(b - a) \end{cases}, \quad f_1 < f_2 \tag{15}$$

$$\begin{cases} a = x_1 \\ x_1 = x_2 \\ x_2 = a + \tau(b - a) \end{cases}, \quad f_1 \geq f_2 \tag{16}$$

The times spent in the turning segments are determined from the solutions to the aforementioned nonlinear systems of equations (Eqs. 6 - 8 or Eqs. 6, 7, and 10). In particular, trust-region-dogleg optimization is used via MATLAB's **`fsolve`**. Trust-region methods use a quadratic model of the objection function and define regions around which the model is trusted to be the proper representation of the objective function [39]. The direction and step length are chosen to either minimize or maximize the function. The context of this discussion will be for a minimization problem. The size of the trust region is tied to the effectiveness of the step size. A trust region too small in size at a particular iteration will slow the algorithm while a trust region that is too large in size may not allow for a feasible step towards the optimal solution in which case the region would have to be reduced.

For a step p (direction and magnitude), iteration k, Hessian approximation B, trust region radius Δ, function f, and model function m, the constrained problem posed for simultaneously finding the step size and direction is defined in Eq. 17.

$$\min\{m(x_k) = f(x_k) + \nabla f(x_k)^T p + \tfrac{1}{2} p^T B(x_k) p\}$$

subject to (17)

$$p^T p \leq \Delta(x_k)^2$$

The dogleg version of trust region optimization finds an approximate solution of the problem posed by Eq. 17 by minimizing the model function along a trajectory which approximates curved paths towards the optimal with linear segments. The path at a particular iteration consists of two line segments. The first segment spans the origin to the current minimum at the particular iteration along the steepest descent direction. The second segment starts from the terminal point of the first and uses the solution for the step, $p^*(\Delta)$ from Eq. 17.

When turn rate and airspeed are coupled, the algorithm searches for a global minimum by solving for the time at points 3 and 6 via **fsolve** and the turn radius in segments 23 and 56 by using interior-point optimization (via MATLAB's **fmincon**). When turn rate and airspeed are coupled, the objective function satisfied by **fmincon** is either the total distance error between the actual and target north-east plane position coordinates or the actual maneuver time with penalty terms that include the error between the actual and desired maximum load factor. The general form of the optimization problem solved by **fmincon** in turning segments is shown in Eqs. 18 and 19 for the distance and time-based modes, respectively. When turn rate and airspeed are decoupled in the time-based mode, there is no need to solve the optimization problem in Eq. 19. Note that R_{c_l} and R_{c_u} are set lower and upper limits on the feasible region in the optimization problem, and k_1 and k_2 are selectable constants.

$$\min\left\{\sqrt{\left(X_{e_{targ}} - X_{e_{act}}\right)^2 + \left(Y_{e_{targ}} - Y_{e_{act}}\right)^2}\right\} \tag{18}$$

subject to

$$R_{c_l} \leq R_c \leq R_{c_u}$$

$$\min\left\{t + k_1\left(N_{z_{max_{targ}}} - N_{z_{max_{act}}}\right)^{k_2}\right\} \tag{19}$$

subject to

$$R_{c_l} \leq R_c \leq R_{c_u}$$

While the global objective is either to minimize an error from target end coordinates of the flight path or minimize time spent in the flight path, each of the eight segments have local objectives that are satisfied using the aforementioned optimization methods. The nonlinear sets of equations that **fsolve** solves involve polynomial representations of path-defining variables (e.g. airspeed, turn rate, altitude). For example, **fsolve** is used to minimize the error between the target and actual airspeeds at points 1 - 8 when solving the airspeed equations for each segment (from Eq. 6) at each of the aforementioned points in time. Another problem is to minimize the error between the integrated turn rate equation (from Eq. 8 or Eqs. 10) for a particular turn segment and the target horizontal flight path angle traversed which is determined using Eqs. 3 - 5. Yet another problem for **fsolve** is to minimize the error between the desired initial and final vertical speeds by comparing the differentiated altitude variation (from Eq. 7) in the straight segments at the initial and final times, to the target vertical speeds.

FLIGHT CONTROL DESIGN

The framework of the flight control design is a nonlinear dynamic inversion control (NLDI) architecture. The plant is based on the nonlinear state-space form given by Eq. 20 with state vector $x(t)$, control input $u(t)$, and output $y(t)$. The inner loop uses full state feedback of only the rigid body states to track the ideal response given by the command filters. The vector function $f(x)$ and matrix function $g(x)$ form the model of the nonlinear aircraft dynamics.

$$\begin{aligned} \dot{x} &= f(x) + g(x)u \\ y &= h(x) \end{aligned} \tag{20}$$

The state vector, which consists of only translation velocities (u,v,w), angular rates (p,q,r), and vehicle attitudes (ϕ,θ) is given in Eq. 21.

$$x = [u \quad v \quad w \quad p \quad q \quad r \quad \phi \quad \theta]^T \tag{21}$$

The forces and moments of the aircraft (FM) can be represented as a linearized expression in terms of perturbations (from trim) in the states and controls along with the redundant controls as given by Eq. 22. SD represents the stability derivative matrix, CD_p the primary control derivative matrix, CD_r the redundant control derivative matrix, and FM_0 the trimmed forces and moments vector. These approximate forces and moments are integrated into the exact equations of motion for a rigid body.

$$FM = SD \cdot \Delta x + CD_p \cdot \Delta u_p + CD_r \cdot \Delta u_r + FM_0 \tag{22}$$

Expressions for $F(x)$ and $G(x)$ can be derived after partitioning the output equation into state dependent and control input dependent terms as given in Eqs. 23.

$$\dot{y} = \frac{dh}{dx}[f(x) + g(x)u] \tag{23}$$

$$F(x) \stackrel{\text{def}}{=} \frac{dh}{dx} f(x)$$

$$G(x) \stackrel{\text{def}}{=} \frac{dh}{dx} g(x)$$

$F(x)$ (feedback linearization vector) and $G^{-1}(x)$ (control mixing matrix) are in turn used to construct the control law given by Eqs. 24 for some feedback compensation $K(s)$; this feedback compensation is either PI or PID. The flight control architecture is given in Fig. 5.

$$\begin{aligned} u &= G^{-1}(x)[\dot{y} - F(x)] \\ &= G^{-1}(x)[\dot{y}_{cmd} - \dot{e} - F(x)] \\ &= G^{-1}(x)[\dot{y}_{cmd} + K(s)e - F(x)] \end{aligned} \tag{24}$$

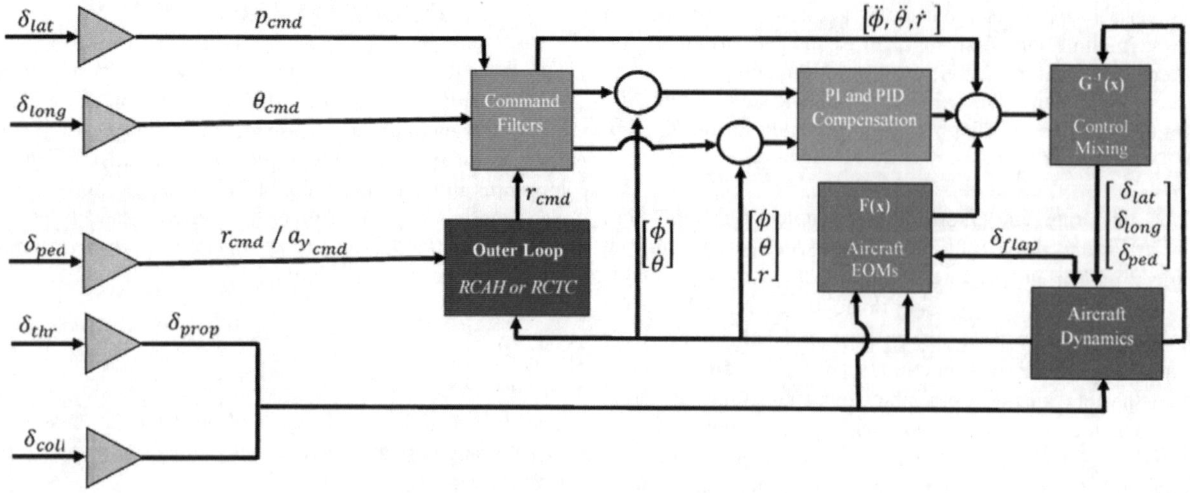

Figure 5. Flight Control Architecture.

BREAK TURN SIMULATIONS

A piloted break turn was performed starting at 161 kts. The pilot performed a 2g, 90 deg. turn with airspeed bleed-off resulting in a final velocity of 151 kts. The FPG was used to give the minimum times required for a 90 deg. turn given a maximum vertical load factor of 2g and the piloted airspeed profile. This ideal time to turn from the FPG can be expressed as a function of the maximum roll rate achieved in the turn as shown in Fig. 6 for a set turn angle, maximum vertical load factor, and airspeed range. The time required in the piloted maneuver based on different maximum roll rates is also marked in Fig. 6. These piloted times are based on the time spent with a non-zero turn rate when turning through 90 deg. but not including the time to capture zero turn rate at the end of the maneuver. Table 2 shows additional break turn simulation times with (actual) piloted time results compared with FPG predictions.

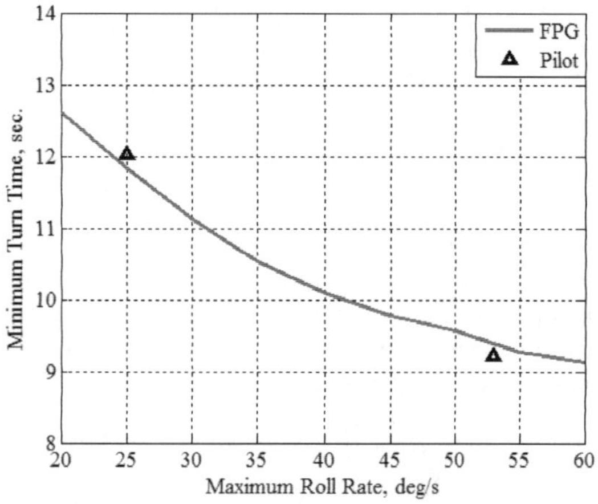

Figure 6. Minimum Times for 2g, 90 deg. Turn.

Note that roll rate and the time rate of change of the bank angle are assumed to be equivalent in the FPG tool; this leads to a source of error. Additional response results are shown in Figs. 7 - 10 for the piloted airspeed range of 161 kts. - 151 kts., maximum vertical load factor of 2g, and maximum roll rate of 25 deg/s. This demonstrates the ability of the mathematical representations of path-defining variables in the FPG to align with actual piloted aircraft responses for airspeed, roll rate, and turn rate in a level turn. Piloted simulations of the break turn successfully discovered handling quality deficiencies. There were a lot of oscillations in roll for high target load factor cases with a tendency for pilot induced oscillations for high roll rate sensitivity values. Lateral cyclic also saturated for high roll rates, and as such, differential wing flaps were allocated to alleviate the roll moment contributions required from the main rotor. While in general the pilot was able to regulate altitude relatively well throughout the maneuver, there was particular difficulty in maintaining airspeed, which to some extent is expected without an automatic airspeed hold.

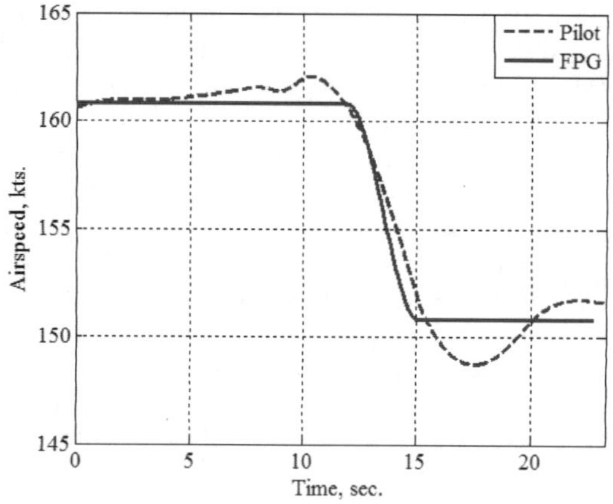

Figure 7. Break Turn - Airspeed Comparison.

Table 2. Break Turn Time Comparisons.

Max Sustained Vertical Load Factor, g	Airspeed Range, kts.	Roll Rate Extrema, deg/s	Pilot Time, sec.	FPG Time, sec.	|Error|, %
1.65	162	-31.5, 35.6	13.07	12.77	2.3
2.12	167 - 161	-29.0, 34.0	11.51	10.83	5.9
2.74	162 - 155	-29.4, 49.5	8.82	8.71	1.2
3.85	167 - 141	-42.0, 35.0	6.64	6.96	4.8
4.67	162 - 130	-41.8, 39.0	5.63	5.26	6.6

Possible quantitative requirements for this MTE may be an overall time requirement along with allowable tolerances for loses in airspeed and altitude to perform the maneuver. The allowable losses in altitude and airspeed can also be combined into one category for an allowable change in specific excess power. Doing so will consequently allow for variations in maneuver; but this may be practical as the maneuver technique to minimize time and energy loss can vary for each aircraft. There is also no loss in generalization when flying at low altitudes as all the energy loss can be put into airspeed, and the allotted power loss constraint due to airspeed and/or altitude loss is a function of gross weight.

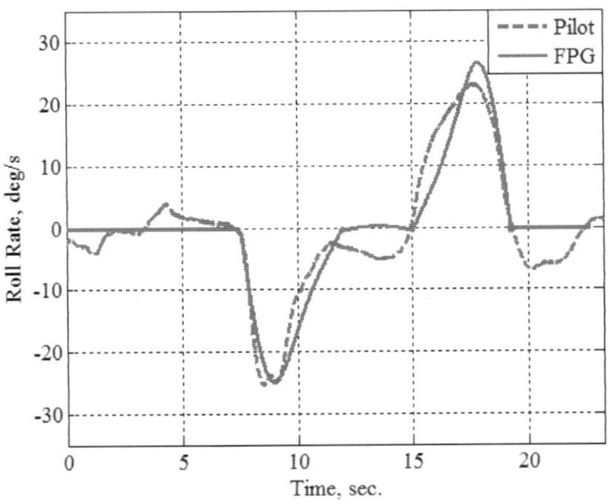

Figure 8. Break Turn - Roll Rate Comparison.

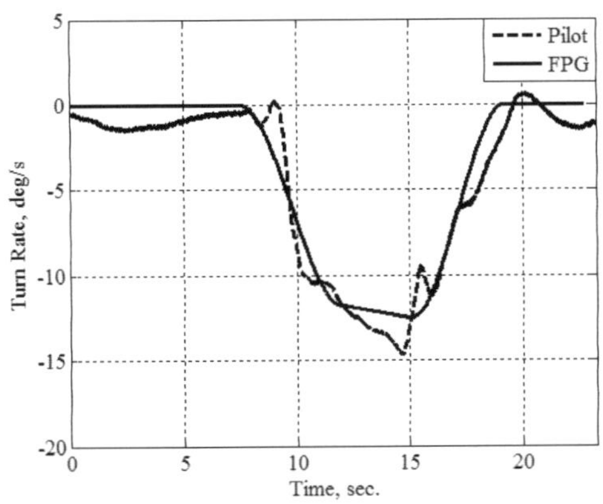

Figure 9. Break Turn - Turn Rate Comparison.

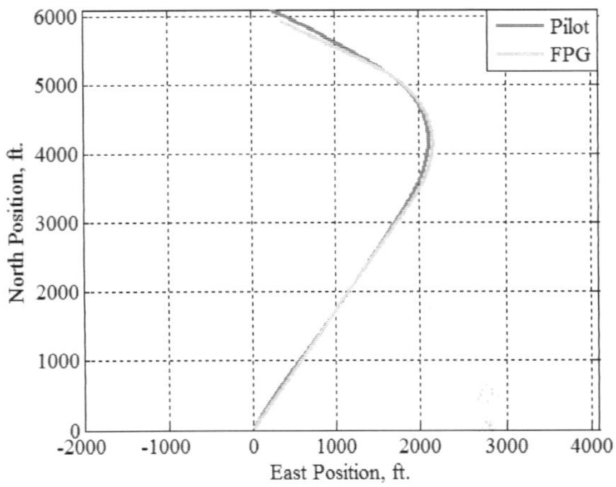

Figure 10. Break Turn - Flight Path Comparison.

DESCENT TURN SIMULATIONS

A descent turn was piloted at 190 kts., starting from 8000 ft., using an advanced controller developed in previous work [31]. A 3.3g descending turn was sustained and then gradually rolled out of, resulting in a varying turn radius. The FPG tool was then used to replicate the piloted flight path of the descent turn. There are multiple ways to replicate the flight path using the FPG tool: the time-based optimization mode with coupled turn rate and airspeed, the decoupled turn rate and airspeed mode, and the distance-based mode. The distance-based mode was used to replicate the flight path (Eqs. 6, 7, 10, and 18). This mode requires the effective turn radius as an input rather than load factor constraints, which were used in the break turn simulations. The set of inputs used to construct the flight path is given in Table 3.

To capture the effect of a varying turn radius, an asymmetric factor (refer to Eqs. 4 and 5) was used. Alternatively another turn segment (refer to Fig. 4) could have been used. A straight-turn-straight flight path was sufficient to replicate this piloted maneuver. Each turn segment is discretized into 3 segments (roll-in, max load factor, roll-out). The asymmetric factor allows for 3 distinct turn radii, which result in a spiraling path. The total change in horizontal flight path angle and effective turn radius was tuned until the flight paths aligned in the north-east plane.

The ground distances covered in the straight segment were set based on the piloted results, along with the altitude change in the first straight segment.

Table 3. Summary of Descent Turn FPG Inputs.

Parameter	Description	Value
dt	time step	0.01 sec.
w_D	distance-based mode flag	1
TRD	decouple turn rate/airspeed flag	0
R_e	effective turn radius	1,150 ft.
χ_0	initial horizontal flight path angle	24.5 deg.
χ_{turn}	total horizontal flight path angle	602 deg.
ξ_T	transient turn factor	0.18
ς	asymmetric turn factor	-0.85
S_0	initial ground distance covered	21,883 ft.
S_f	final ground distance covered	1,494 ft.
h_0	initial altitude	8,000 ft.
h_1	altitude at point 1	5,400 ft.
$V_{1,2\ldots5}$	airspeed at points 1 - 5	191 kts.
ΔN_{z_1}	change in load factor at point 1	-0.05g
ΔN_{z_2}	change in load factor at point 2	-0.025g
ΔN_{z_3}	change in load factor at point 3	0.080g
ΔN_{z_4}	change in load factor at point 4	0.135g
ΔN_{z_5}	change in load factor at point 5	0g

The comparison of the flight paths is shown in Fig. 11. The projection of the piloted flight path onto the north-east plane is also shown in Fig. 11. The changes in vertical load factor along with flight path were the last inputs that were tuned to match the changes in altitude over the course of the entire maneuver.

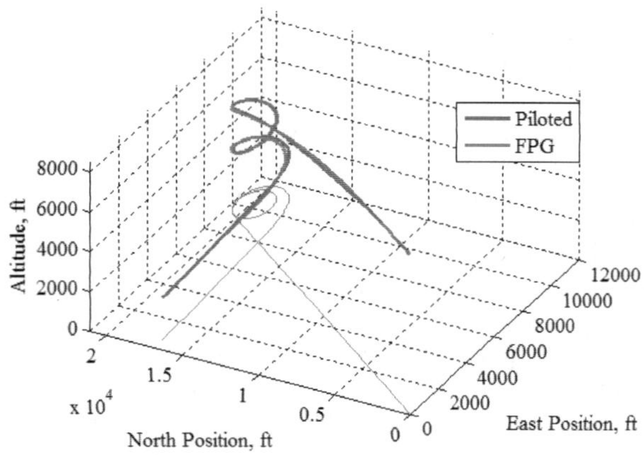

Figure 11. Descent Turn - Flight Path Comparison.

CONCLUSIONS

1. Results from the FPG predict break turn maneuver times that align with the piloted times given a target maximum vertical load factor and maximum roll rate over a set airspeed range. The FPG tool was able to predict break turn times with errors ranging from 2% - 7% across a variety of airspeed ranges, load factors, and roll rate extrema.

2. The controller filter and compensator gains need to be more rigorously tuned as not much work with this has been performed. In particular, the lateral-axis controller needs work to remedy oscillatory behavior at high vertical load factors. This deficiency was exposed by the series of piloted break turns performed in this study.

3. The FPG tool is able to accurately replicate flight paths of generalized maneuvers, as demonstrated with the descent turn. This maneuver features straight and turning segments of varying altitude and turn radii.

FUTURE WORK

1. A decelerating approach should be examined as an MTE. Work on this MTE is currently underway using a PAPI approach setup, but specific time constraints and target glideslopes are still being investigated.

2. Specific performance metrics and constraints for the break turn such as allowable altitude and/or airspeed changes and time constraints should be developed to characterize this MTE.

3. Specific performance metrics and constraints for the descent turn should be developed, such as a target change in horizontal flight path angle, maximum load factor, target altitude change, and a time constraint.

ACKNOWLEDGEMENTS

The U.S. Army's National Rotorcraft Technology Center program is cofounding this research effort sponsored by the U.S. Government under transaction number W15QKN-10-9-0003 between Vertical Lift Consortium, Inc. and the Government. The U.S. Government is authorized to reproduce and distribute reprints for Governmental purposes notwithstanding any copyright notation thereon. The views and conclusions contained herein are those of the authors and should not be interpreted as necessarily representing the official policies or endorsements, either expressed or implied, of the U.S. Government.

REFERENCES

[1]Sekula, M. K., and Gandhi, F., "Effects of Auxiliary Lift and Propulsion on Helicopter Vibration Reduction and Trim", *Journal of Aircraft*, Vol. 41, No. 3, May 2004, pp. 645-656.

[2]Gandhi, F., and Sekula, M. K., "Helicopter Vibration Reduction using Fixed-System Auxiliary Moments", *Journal of Aircraft*, Vol. 41, No. 3, March 2004, pp. 645-656.

[3]Horn, J. F., Guo, W., and Ozdemir, G.T., "Use of Rotor State Feedback to Improve Closed-Loop Stability and Handling Qualities", *American Helicopter Society 66th Annual Forum*, 2010.

[4]Chen, R., "An Exploratory Investigation of the Flight Dynamics Effects of Rotor RPM Variations and Rotor State Feedback in Hover", NASA-TM-103968, NASA Ames Research Center, Sept. 1992.

[5]Steiner J. and Farhan G., "An Investigation of Variable Rotor RPM on Performance and Trim", *American Helicopter Society 64th Annual Forum*, Montreal, Canada, April-May 2008.

[6]Horn, J.F. and Guo, W., "Flight Control Design for Rotorcraft with Variable Rotor Speed", *American Helicopter Society 64th Annual Forum*, Montreal, Canada, April-May 2008.

[7]Horn, J.F. and Guo, W., "Helicopter Flight Control with Variable Rotor Speed and Torque Limiting", *American Helicopter Society 65th Annual Forum*, Grapevine, TX, May 2009.

[8]Geiger, B.R., "Flight Control Optimization on a Fully Compounded Helicopter with Redundant Control Effectors", M.S. Thesis, Dept. of Aerospace Engineering, The Pennsylvania State University, University Park, PA, May 2005.

[9]Piasecki et al. "Compound Aircraft Control System and Method", United States Patent. No.: US 8,463,465 B2. Jun. 11, 2013.

[10]Ozdemir, G.T. and Horn, J. F., "Simulation Analysis of a Flight Control Law with In-Flight Performance Optimization", *American Helicopter Society 68th Annual Forum*, Fort Worth, Texas, May 1-3, 2012.

[11]Gilyard, G., "In-flight Adaptive Performance Optimization (APO) Control Using Redundant Control Effectors of an Aircraft", U.S. Patent 5,908,176, June 1, 1999.

[12]Gilyard, G., "In-flight Transport Performance Optimization: An Experimental Flight Research Program and an Operational Scenario." *AIAA/IEEE 16th Digital Avionics Systems Conference*, Irvine, CA, Oct. 1997, pp. 7.2.33-7.2.49.

[13]Gilyard, G., "Development of a Real-Time Transport Performance Optimization Methodology", NASA-TM-4730, NASA Dryden Flight Research Center, Jan. 1996.

[14]Gilyard, G., "Flight Test of an Adaptive Configuration Optimization System for Transport Aircraft", NASA-TM-1999-206569, NASA Dryden Flight Research Center, Jan. 1999.

[15]Ozdemir, G.T., Horn, J.F., Thorsen, A.T., "In-Flight Multi-Variable Optimization of Redundant Controls on a Compound Rotorcraft", *AIAA Guidance, Navigation, and Control Conference*, August 2013.

[16]Ozdemir, G.T., "In-Flight Performance Optimization for Rotorcraft with Redundant Controls." Ph.D. Dissertation, Dept. of Aerospace Engineering, The Pennsylvania State University, University Park, PA, December 2013.

[17]Thorsen, A.T., Horn, J.F., Ozdemir, G.T., "Optimizing Control of a Compound Rotorcraft in Quasi-Steady Maneuvers", *Fifth Decennial AHS Aeromechanics Specialists' Conference,* San Francisco, CA, January 2014.

[18]Thorsen, A.T., "Assessment of Control Allocation Optimization on Performance and Dynamic Response Enhancement of a Compound Rotorcraft", M.S. Thesis, Dept. of Aerospace Engineering, Pennsylvania State University, University Park, PA, Aug. 2014.

[19]Thorsen, A.T., Horn, J.F., Ozdemir, G.T., "Use of Redundant Controls to Enhance Transient Response and Handling Qualities of a Compound Rotorcraft", *American Helicopter Society 70th Annual Forum,* Montreal, Quebec, May 20-22, 2014.

[20] Pavel, M.D., Padfield, G.D., "Progress in the Development of Complementary Handling and Loading Metrics for ADS-33 Manoeuvres", *Proceedings of the American Helicopter Society 59th Annual Forum, 2003.*

[21] Soneson, G.L., Horn, J.F., "Simulation Testing of Advanced Response Types for Ship-Based Rotorcraft", *American Helicopter Society 70th Annual Forum,* Montreal, Quebec, May 20-22, 2014.

[22] Soneson, G.L., "Development and Piloted Simulation Testing of Advanced Response Types for Ship-based Rotorcraft", M.S. Thesis, Dept. of Aerospace Engineering, Pennsylvania State University, University Park, PA, Aug. 2014.

[23] Zheng, A., Horn, J.F., "Investigation of Bandwidth and Disturbance Rejection Properties of a Dynamic Inversion Control Law for Ship-Based Rotorcraft", *American Helicopter Society 71st Annual Forum,* Virginia Beach, VA, May 5-7, 2015.

[24] Zheng, A., "Investigation of Bandwidth and Disturbance Rejection Properties of a Dynamic Inversion Control Law for Ship-based Rotorcraft", M.S. Thesis, Dept. of Aerospace Engineering, Pennsylvania State University, University Park, PA, Dec. 2015.

[25] Yomchinda, T., Horn, J.F., Langelaan, J.W., "Flight Path Planning for Descent-phase Helicopter Autorotation", *AIAA Guidance, Navigation, and Control Conference*, Portland, Oregon August 8-11 2011.

[26] Yomchinda, T., Horn, J.F., Langelaan, J.W., "Modified Dubins parameterization for aircraft emergency trajectory planning", *Journal of Aerospace Engineering, Institution of Mechanical Engineers* Part G, 2016.

[27] Cao, Y., *Principles of Helicopter Flight Dynamics*, Coxmoor Publishing Company, Oxford, UK, 2009, pg. 191-230.

[28] Ferguson, K., Thomson, D., "Maneuverability Assessment of a Compound Helicopter Configuration", *Journal of the American Helicopter Society*, June 2015.

[29] Thorsen, A.T., Horn, J.F., "Aerobatic Maneuvers of a Conventional and Compound Rotorcraft - Flight Path Generation and Dynamic Simulation", *AIAA Aviation 2015 Atmospheric Flight Mechanics*, Dallas, TX, June 2015.

[30] Thorsen, A.T., Horn, J.F., "Aerobatic Maneuvers of a Compound Rotorcraft at High Advance Ratios - Flight Path Generation and Dynamic Simulation", *AIAA Aviation 2016 Atmospheric Flight Mechanics*, Washington, D.C., June 2016.

[31] Thorsen, A.T., "A Unified Flight Control Methodology for a Compound Rotorcraft in Fundamental and Aerobatic Maneuvering Flight", Ph.D. Dissertation, Dept. of Aerospace Engineering, Pennsylvania State University, University Park, PA, Dec. 2016.

[32] King, R. and Wiggins, L., "Draft Changes to ADS-33E-PRF", U.S. Army AED - Redstone Arsenal, Aviation and Missile Research Development and Engineering Center, April 23, 2010.

[33] Roncoli, R.B., "A Flight Test Maneuver Autopilot for a Highly Maneuverable Aircraft", NASA Technical Memorandum 81372, Nov. 1982.

[34] Dunford, P.J., Marr, R.L., Martin, Col. Paul, Price, Lt. Col. Bob, "V-22 Flight Test Update", *European Rotorcraft Forum*, Brighton, UK, Sept. 1996.

[35] Howlett, J., *UH–60A BLACK HAWK Engineering Simulation Program: Volume I –Mathematical Model*, NASA CR–166309, Dec. 1981.

[36] Horn, J. F., Bridges, D. O., Lopes, L. V., and Brentner, K. S., "Development of a Low-Cost, Multi-Disciplinary Rotorcraft Simulation Facility", *Journal of Aerospace Computing, Information and Communication,* Vol. 2, No. 7, July 2005, pp. 267-284.

[37] McCormick, B.W., *Aerodynamics, Aeronautics, and Flight Mechanics.* John Wiley & Sons Inc., 1994, pgs. 291-316.

[38] Belegundu, A.D. and Chandrupatla, T.R., *Optimization Concepts and Applications in Engineering.* Cambridge University Press, 2011, pgs. 63-67.

[39] Nocedal, J. and Wright, S.J., *Numerical Optimization.* Springer Science+Business Media, 2006, pgs. 66-76, 480-485.

MH-60M Directional Control Margin Investigation

Jeffrey Von Hor
Experimental Test Pilot
US Army, Redstone Test Center
Redstone Arsenal, AL USA

Allan Guess
Flight Test Engineer
US Army, Redstone Test Center
Redstone Arsenal, AL USA

ABSTRACT

A directional control margin investigation of the MH-60M Black Hawk helicopter was conducted to address the loss of directional control when operating at high density altitudes and high gross weights. This paper will discuss the flight test techniques used to assess directional control margin, the process for determining gross weight for flight testing at high density altitude, and resolution of the loss of directional control on the MH-60M. Additionally, low density altitude, low gross weight, and low power flight tests were performed to ensure adequate right directional control margin remained after implementation of changes to improve the loss of directional control at high density altitudes and high gross weights. A relationship between power required and directional control margin was established. Based on that relationship, a flight test methodology was developed to successfully quantify the directional control margin (DCM) of the MH-60M in stabilized hover and in the low airspeed flight regimes.

NOTATION

$$C_P = \frac{P}{\rho_a A_D (\Omega R)^3}$$

$$C_T = \frac{T}{\rho_a A_D (\Omega R)^2}$$

$$W_{REF} = \frac{W_T}{\sigma_T}$$

A_D	rotor disk area
AGL	above ground level
C_P	coefficient of power
C_T	coefficient of thrust
DCM	directional control margin
ft	feet
kts	knots
KTAS	knots true airspeed
lbs	pounds
N_R	main rotor speed
P	power
SHP	shaft horsepower
TRIMPP	tail rotor impressed pitch
W_{REF}	referred gross weight
W_T	test gross weight
ρ_a	ambient air density
σ_T	test density ratio
ΩR	blade rotational velocity

INTRODUCTION

Aircraft growth creep is common for many legacy airframes like the UH-60 Black Hawk. Initially developed in the mid-1970s, the Utility Tactical Transport Aircraft System (UTTAS), later designated the YUH-60A, had a mission gross weight of 16,750 lbs[1]. Today, the newest model of the Black Hawk helicopter in the US Army, the UH-60M, has a maximum takeoff gross weight of 22,000 lbs[2]. The MH-60M, a modified UH-60M, further increases the maximum takeoff gross weight of the Black Hawk to 24,500 lbs[3]. Over the decades, operational requirements for the H-60 airframe have continued to evolve requiring modifications to the airframe, transmission, engines, and rotor system to meet those operational requirements.

The development of the SH-60B Seahawk for the US Navy in 1983 introduced an improved durability gearbox rated for 3,400 horsepower, increased from 2,828 horsepower on the UH-60A gearbox. The Seahawk had uprated T700-GE-701C engines which increased intermediate rated power to 1,857 shaft horsepower, increased from 1,622 shaft horsepower on the T700-GE-700 engines on a UH-60A[4]. Though the performance of the Seahawk and Black Hawk helicopters was improved with the introduction of the improved durability gearbox and uprated engines, the main rotor and tail rotor airfoils remained unchanged. During the 1990s, the tail rotor bias (the offset from the neutral pitch setting of the tail rotor blade) was increased from 1.5 degrees to 3.0 degrees to increase tail rotor authority at higher gross weights at high density altitudes[5]. With fielding of the UH-60M in 2007, a new and modern main rotor blade airfoil was introduced to the Black Hawk. The chord of the UH-60M main rotor blade was increased to 24.25 inches (16 percent longer than the UH-60A/L main rotor blade) with a SC2110 airfoil and an aft swept anhedral tip with a SSCA09 airfoil[6]. However, the tail rotor airfoil and tail rotor pitch bias remained unchanged. During flight testing of the MH-60M in the late 2000s and early 2010s, a requirement to address tail rotor authority at high gross weights and high density altitudes became apparent.

The MH-60M

The MH-60M Black Hawk helicopter (figure 1) is a UH-60M modified by the Lockheed Martin Corporation to support the Special Operations Aviation Regiment. In addition to structural modifications to support a maximum takeoff gross weight of 24,500 lbs, the T700-GE-701D engines of the UH-60M (intermediate rated power of 1,902 shaft horsepower) were replaced with YT706-GE-700 engines (intermediate rated power of 2,567 shaft horsepower[7]). However, the main transmission, main rotor and tail rotor components are identical to those on the UH-60M. The increased performance of the MH-60M with the YT706-GE-700 engines coupled with the increased maximum gross weight resulted in directional control issues during low airspeed, high gross weight, high density altitude testing.

Figure 1. MH-60M Black Hawk Helicopter

FLIGHT TESTING

Loss of Tail Rotor Authority

The first high gross weight, high density altitude test of the MH-60M was conducted in Alamosa, CO (elevation 7,539 ft) in August – September 2010. The purpose of the testing in Alamosa was to collect high altitude hover performance data and to conduct a flight loads survey, not to conduct a handling qualities evaluation. The aircraft was configured for flight loads survey during low airspeed flight at a main rotor speed of 100 percent, with no external stores, and with aircraft internally ballasted to achieve the maximum gross weight for an in-ground-effect hover minus a torque margin. Low airspeed testing was flown in winds less than 5 kts. The tail rotor bias was set to 3 degrees which resulted in a tail rotor impressed pitch (TRIMPP) range of -12.5 degrees (full right pedal, full down collective) to 18.5 degrees (full left pedal, full up collective). Standard flight test practice is to conduct an engineering flight control rigging check which includes measuring the actual maximum TRIMPP by placing a digital protractor on the tail rotor blade cuff with full left pedal and full up collective. During the engineering flight control rigging check, the measured maximum TRIMPP angle was 18.8 degrees. The 0.3 degree delta between the maximum TRIMPP of 18.5 degrees and the measured 18.8 degrees maximum TRIMPP was due to the tail rotor control rigging procedure. This rigging inaccuracy led to operational planning difficulties which will be discussed later in this paper.

During the flight loads survey test in Alamosa, CO, the loss of directional control of the MH-60M manifested itself as a slow right rotation as the aircraft passed through the translational lift region – the low airspeed regime between 10 to 20 knots true airspeed (KTAS). The loss of directional control was experienced on the 045, 090, 135, 180 and 225 degree azimuths. As the right rotation was controllable with a reduction on collective or increasing forward speed, the test team assessed the tail rotor was not stalled. For experienced helicopter pilots who fly helicopters with a tail rotor, loss of directional control is not uncommon. Loss of tail rotor effectiveness, loss of tail rotor thrust, and left yaw due to an engine failure are topics of discussion for many student helicopter pilots especially if they are learning to fly on single-engine, light-weight helicopters with a tail rotor. So, why was the loss of directional control in the Black Hawk an issue?

Historically, the Black Hawk has been engine and/or transmission limited, not tail rotor limited. Prior to the MH-60M, the typical environment where a Black Hawk pilot could experience a loss of directional control would be while attempting to land or hover in a high temperature environment, at high altitude, and at high gross weight – the proverbial high, hot and heavy environment. In these conditions, the engine reaches the operating temperature limit and the fuel control unit begins to limit fuel flow while the pilot continues to increase collective to arrest a rate of descent on an approach or to stabilize in a hover. As the pilot continues to increase collective, power required exceeds power available and the rotor speed begins to droop (slow) which slows the tail rotor speed reducing the thrust produced by the tail rotor. The anti-torque thrust of the tail rotor is no longer sufficient to counter the torque applied to the helicopter fuselage and the helicopter begins to spin to the right. For the MH-60M, the loss of tail rotor control was not due to engine limiting, as the new YT706-GE-700 engines produced plenty of power, but a yaw control margin and tail rotor impressed pitch limitation.

Tail Rotor Impressed Pitch vs. Yaw Control Margin

One of the distinguishing features of the Black Hawk helicopter is the canted tail rotor. The tail rotor is mounted on the right-side of the vertical empennage. When viewed from the back of the aircraft, the tail rotor is canted 20 degrees counter-clockwise. The tail rotor is a tractor type tail rotor which means the lift vector is pulling the tail of the aircraft to the right. The canted tail rotor introduces inherent coupling with power changes (collective control increase/decrease) or directional control movement. There are five different tail rotor coupling modes: collective to yaw, collective to pitch, collective to roll, yaw to pitch, and collective/airspeed to yaw.

Four of the modes are reduced through the mechanical mixing unit and one coupling mode is reduced through the flight control computer via the trim system[8]. The collective to yaw coupling mode and the compensation applied by the mechanical mixing unit are of the most interest in the low airspeed regime.

A helicopter flight control system, whether purely mechanical or hydraulically assisted, has flight control margins (control envelopes) for the collective, cyclic and directional control pedals that are limited by physical stops. This is also true of the Black Hawk flight control system with one caveat. Though the directional control pedals have physical stops that define the boundary of the control envelope, the mechanical mixing unit increases pitch to the tail rotor blades as the collective is increased to compensate for the collective to yaw coupling. The additional tail rotor pitch being applied by the mechanical mixing unit does not physically move the yaw control pedals (inner-loop) which results in flight regimes where the tail rotor blade is at its maximum impressed pitch angle but the left yaw pedal is not on the control stop. As the pilot continues to push on the left pedal, the left pedal will continue to move forward until the control stop is reached but no additional pitch will be applied to the tail rotor blades. For low airspeed testing of directional control, monitoring tail rotor impressed pitch, measured at the tail rotor pitch change shaft, was as important as monitoring the yaw pedal control margin.

Directional Control Margin

Having two separate directional control limitations, tail rotor impressed pitch and yaw pedal control margin, complicated the reporting of the loss of directional control during high density altitude, high gross weight testing. Classically for a low airspeed handling qualities evaluations, plots of the flight control positions at incremental speeds from a hover to the maximum sideward/rearward flight speed are plotted for each relative wind azimuth tested. Directional control position vs. KTAS for the 045 degree relative wind azimuth with error bars representing maximum/minimum directional control movement is presented in figure 2. When a flight control has less than 10 percent margin remaining, that flight condition is highlighted as an area of concern. However, for the Black Hawk, having a 10 percent yaw pedal control margin remaining didn't necessarily assure there was an additional 10 percent of tail rotor impressed pitch remaining. To capture this disparity, the term direction control margin was used. Directional control margin being defined as the lesser of the tail rotor impressed pitch and yaw pedal control margin remaining.

Defining the Directional Control Problem

The year after the identification of the loss of directional control in Alamosa, CO, a full low airspeed handling qualities evaluation of the MH-60M was made. Initially, the low airspeed handling qualities evaluation was made at low density altitude, at high gross weight in Lexington, KY (elevation 947 ft). The MH-60M was configured at the maximum takeoff gross weight and testing was conducted in winds less than 5 kts using a pace vehicle. Eight relative wind azimuths were tested (the cardinal and ordinal azimuths) from a hover to 45 KTAS in 5 knot increments. During the low airspeed testing in Lexington, KY, no directional control margin issues were identified.

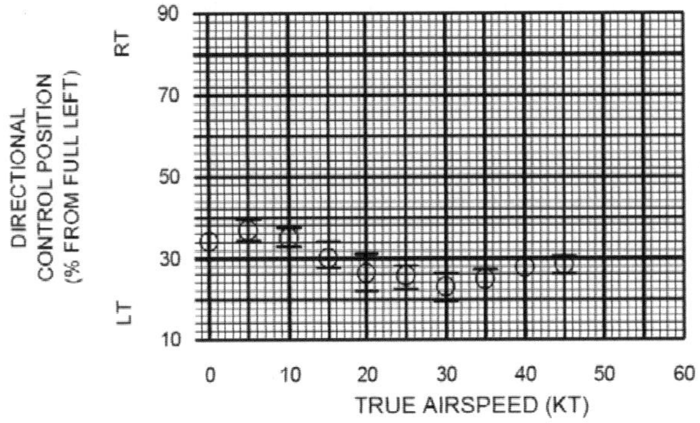

Figure 2. Directional Control Position vs. KTAS
(045° relative wind azimuth)

After completing low airspeed handling qualities testing in Lexington, KY, the test team moved to Colorado to conduct low airspeed, high density altitude, high gross weight testing at Buena Vista, CO (elevation 7,950 ft) and Leadville, CO (elevation 9,933 ft) from July to August 2011. The purpose of the low airspeed, high density altitude testing was to determine the maximum gross weight that directional control could be maintained on all eight relative wind azimuths from the hover to 45 KTAS. After several weeks of testing in both Buena Vista and Leadville, the critical wind speeds and azimuths were identified based on directional control margin. The data are presented in the form of a critical azimuth plot in figure 3. As the Black Hawk typically does not have adverse handling qualities in the low airspeed regime, the critical wind speeds and azimuths in figure 3 were defined as having less than 10 percent directional control margin remaining. From figure 3, the critical azimuths are the relative wind azimuths from the 045 degree azimuth clockwise around to the 270 degree azimuth. The 045 and 090 degree relative wind azimuths are critical from 5 kts to 45 kts and have regions with less than 5 percent directional control margin remaining.

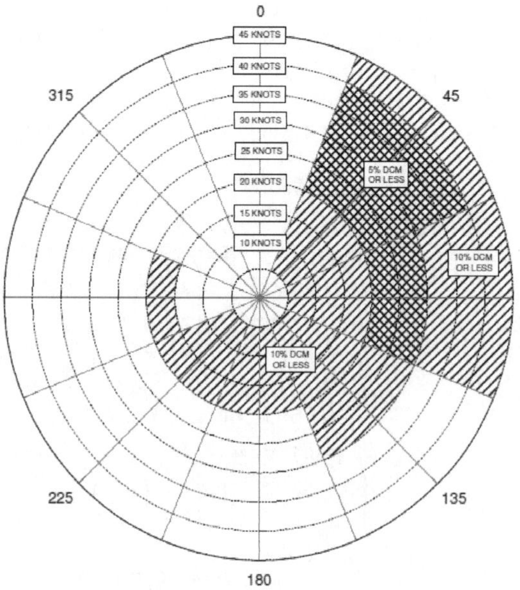

Figure 3. MH-60M Directional Control Margin
(20 ft AGL, 100% N_R, 18.5° Impressed Pitch)

Coefficient of Power is the Key

As stated in the previous paragraph, low airspeed, high density altitude testing was flown at both Buena Vista and Leadville, CO. So, why is there only one critical azimuth plot presented? Initially, two separate critical azimuth plots were made. However, the azimuth plots were almost identical, differing only at one speed on the 045 degree relative azimuth. After reviewing the flight test data for that test point, that difference was attributed to the time slice used to pull data for control positions. An engineering judgment was made that the critical azimuth and speed data from Leadville matched the critical azimuth and speed data recorded at Buena Vista, CO. The test team then sought to understand how the critical azimuth and speed data collected at two different airports separated by 2,000 ft of altitude, flown at two different test gross weights, on two different test days could be the same.

The flight test engineers processed the data for the low airspeed tests flown at Buena Vista and Leadville, CO in addition to the tethered hover performance data flown at both locations, Alamosa and Lexington, KY. Various plots of the tethered hover data were made looking for a relationship between two sets of data. Data plots of: gross weight vs. TRIMPP (figure 4), collective control position vs. TRIMPP (figure 5), engine torque vs. TRIMPP (figure 6), referred gross weight vs. TRIMPP (figure 7), coefficient of thrust vs. TRIMPP (figure 8), and coefficient of power vs. TRIMPP (figure 9) were made and evaluated. The data plotted in figures 4 through 9 are from tethered hover performance flights collected at multiple density altitudes and at three rotor speeds (95, 100 and 105 percent) at each location. As can be seen from figures 4 through 8, there is no single relationship between the sets of data plotted. However, in figure 9, there is a nearly linear relationship between coefficient of power (the non-dimensional power coefficient) and TRIMPP. With a relationship between the coefficient of power and TRIMPP identified for stable hovering flight, the test team began searching for a way to apply that relationship in a broader context of low airspeed flight.

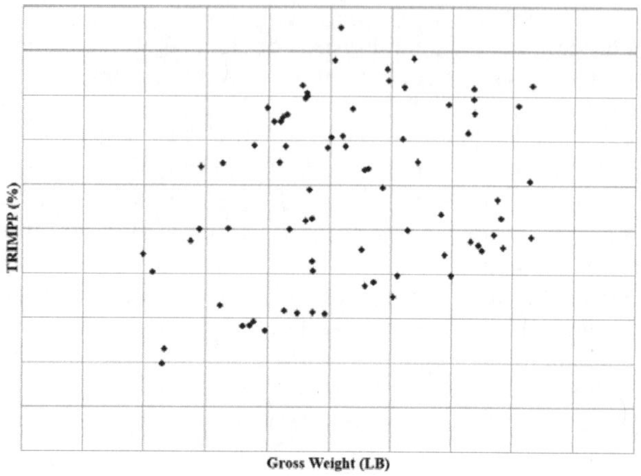

Figure 4. Gross Weight vs. TRIMPP

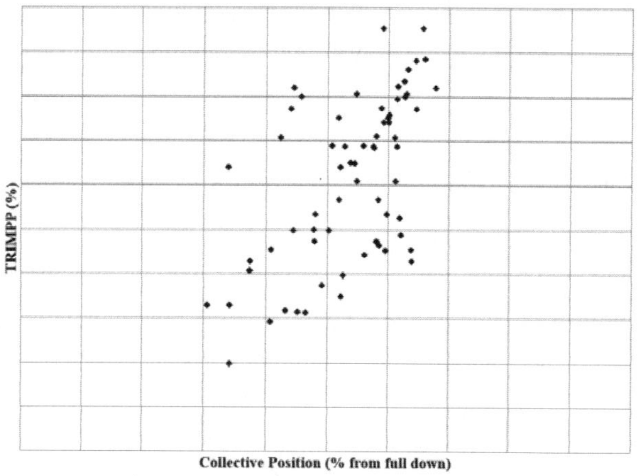

Figure 5. Collective Control Position vs. TRIMPP

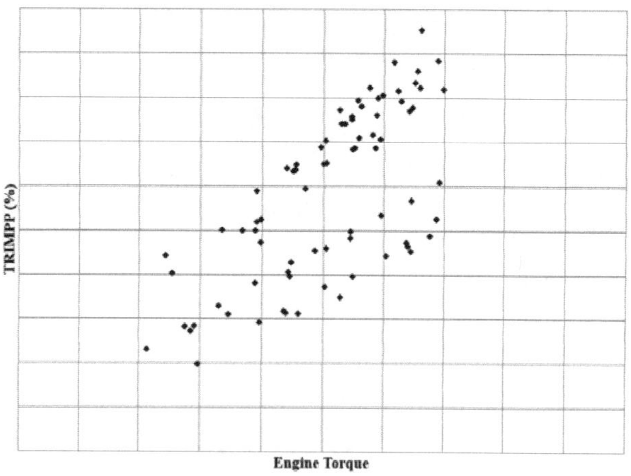

Figure 6. Engine Torque vs. TRIMPP

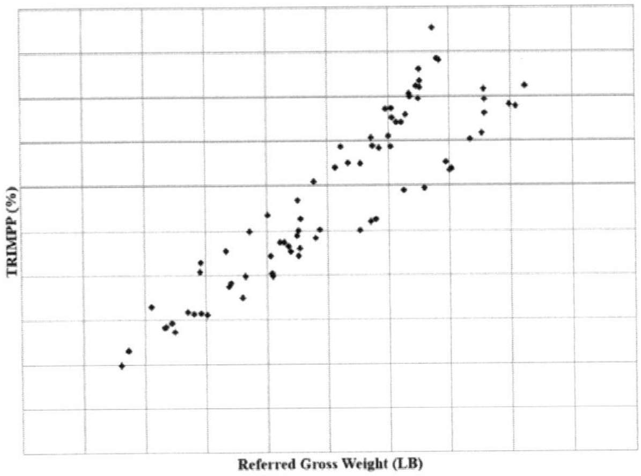

Figure 7. Referred Gross Weight vs. TRIMPP

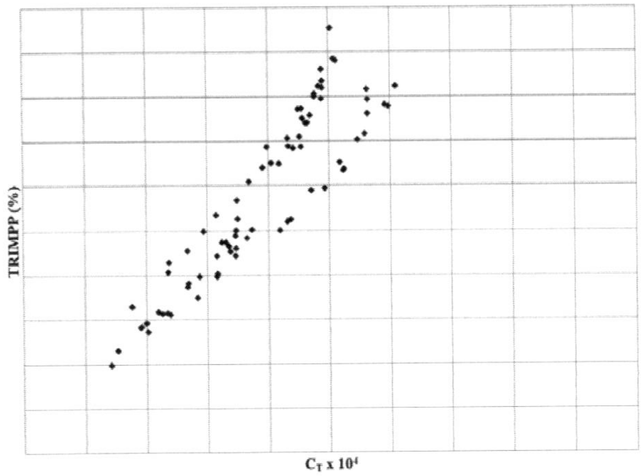

Figure 8. C_T x 10^4 vs. TRIMPP

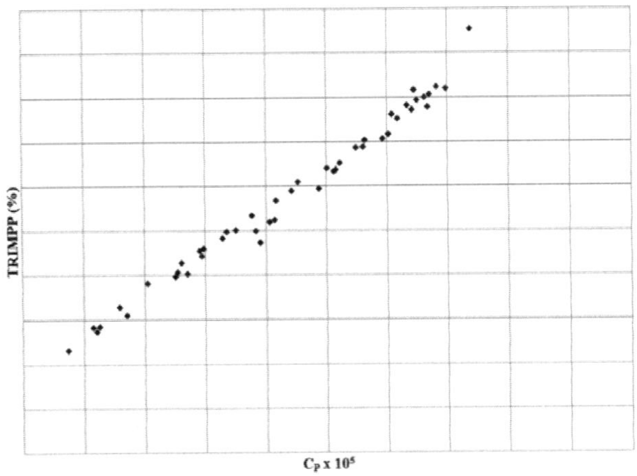

Figure 9. C_P x 10^5 vs. TRIMPP

Unfortunately, there was no direct means to apply the coefficient of power and TRIMPP relationship to low airspeed flight. The coefficient of power vs. TRIMPP relationship identified in figure 9 was for a static flight condition (hover) in winds less than 3 kts. Low airspeed flight conditions are not static flight conditions and data are not collected in winds less than 3 kts. However, the initial data point flown during low airspeed flight test is a hover point. That hover point was the key to determining the initial referred gross weight to start low airspeed testing. Based on previous MH-60M low airspeed flight test data, the test team identified the relative wind azimuth and speed with the minimum tail rotor impressed pitch remaining. The minimum TRIMPP remaining was compared to the TRIMPP remaining at the hover point. The delta between the TRIMPP in a hover and the minimum TRIMPP was determined. That delta was added to the hover point TRIMPP to establish the target TRIMPP. Using the coefficient of power and TRIMPP relationship, the test team was able to calculate the hover power required to achieve the target TRIMPP. From MH-60M hover performance data, the test team was able to estimate the aircraft gross weight that corresponded to the hover power required based on the test day conditions. In theory, using the relationship between coefficient of power and TRIMPP, determining a delta TRIMPP, adding that delta to the hover point TRIMPP, then determining hover power required, and finally estimating the actual aircraft hover gross weight would work. But, did it work in practice.

From Theory to Practice

The opportunity to put the theory into practice presented itself in 2012 when the test team returned to Buena Vista, CO for additional high density altitude flight testing. This trip to Buena Vista was specifically to conduct directional control margin testing. Based on the Colorado flight testing in 2011, the test team knew there was no requirement to test at both Buena Vista and Leadville. For the 2012 Buena Vista testing, the MH-60M users gave the test team a specific requirement. The object of the directional control margin test was to determine the maximum gross weight the aircraft could be flown while maintaining a 10 percent directional control margin on all 8 relative wind azimuths out to 10 KTAS for both in-ground-effect and out-of-ground-effect flight conditions. The test configuration remained no external stores, 100 percent rotor speed (N_R), 3 degrees of tail rotor bias (18.8 degrees maximum TRIMPP). Additionally, an investigation of the benefit of using 5 degrees of tail rotor bias (20.4 degrees maximum TRIMPP) was performed. Using the coefficient of power vs. TRIMPP relationship determined following the 2011 low airspeed testing, the test team was able to determine the maximum initial test aircraft gross weight that would maintain a 10 percent directional control margin out to 10 KTAS on all 8 relative azimuths within approximately 200 lbs. referred gross weight.

NEXT STEPS

After completing the directional control margin testing at 100 percent N_R in-ground-effect and out-of-ground-effect, the test team flew a reduced scope exploration of the effects on directional control margin when the aircraft was flown with 20.4 degrees maximum TRIMPP in-ground-effect only. Tests were flown on all 8 relative wind azimuths from the hover out to the maximum sideward/rearward flight speed. Test objectives remained the same, so gross weight was increased to achieve a minimum of 10 percent directional control margin out to 10 KTAS on all 8 relative wind azimuths. The results of that testing indicated a performance increase of approximately 1,500 lbs referred gross weight with the increase in TRIMPP to 20.4 degrees. Figure 10 is a plot of directional control margin at 20 ft. above ground level (AGL) at both 18.8 and 20.4 degrees TRIMPP. The plots for both tail rotor impressed pitch angles are the same as the objectives of the test were the same and the relationship between coefficient of power and TRIMPP proved to be correct. The nearly linear relationship between the coefficient of power and tail rotor impressed pitch is only valid along the linear portion of the coefficient of lift vs. alpha (angle of attack) curve. Once the tail rotor enters the stall region of the curve, the linear relationship is no longer valid.

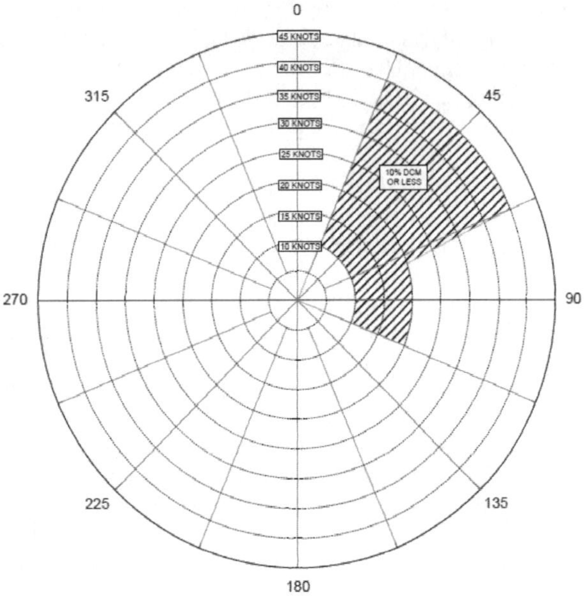

Figure 10. MH-60M Directional Control Margin
(20 ft AGL, 100% NR, 18.8° and 20.4° Impressed Pitch)

The data from the 2012 Buena Vista directional control margin testing was presented to the user later the following year along with the results of the testing at 20.4 degrees TRIMPP. The user was very interested in the increased performance at high density altitudes that was indicated by the data from the 20.4 degrees TRIMPP testing. Additionally, the users, who were training on the MH-60M by this time, reported an issue with the performance charts regarding directional control margin. During high altitude training, a three-aircraft formation would make an approach to a hover. Two of the aircraft would enter an out-of-ground-effect hover with no issue but the other aircraft would lose directional control and have to abort the approach. The user reported the planning charts, which depicted direction control margin based on density altitude and aircraft gross weight, did not provide any consistency and were not satisfactory as a planning tool.

Optimized Tail Rotor Rig

Following the meeting with the users, the test team set out to determine why the performance planning charts were not accurately predicting directional control margin. One of the paths investigated was to check the tail rotor bias and tail rotor impressed pitch on the operational aircraft. As part of flight test, engineering rigging checks of the flight controls are made which documents control envelopes, control forces, main rotor and tail rotor blade angles at various collective and yaw pedal positions. When the maximum tail rotor impressed pitch of the operational aircraft was checked using a digital protractor, a wide variation in the maximum pitch angle was discovered. The performance planning charts the users were using were developed from flight test data based on a test aircraft with a maximum TRIMPP of 18.5 degrees which is the prescribed maximum pitch for a tail rotor with 3 degrees of bias. The operational aircraft surveyed had maximum TRIMPP angles varying from below 17.5 degrees to angles in excess of 21 degrees. The tail rotor rigging procedure in use at the time was not providing consistency in the maximum TRIMPP; therefore, the planning charts would not accurately predict directional control margin. An improved tail rotor rigging procedure was needed that would accurately set the maximum TRIMPP.

An improved tail rotor rigging procedure was developed which was called the optimized tail rotor rig (OTR). Without getting into the details of rigging a tail rotor, the optimized tail rotor rig procedure set the maximum blade angle using a digital protractor to the desired blade angle. Therefore, all the MH-60M aircraft would have the same maximum TRIMPP set and the planning charts would accurately predict directional control margin. Based on the directional control margin testing at Buena Vista in 2012 with 20.4 degrees of maximum TRIMPP, the user requested the tail rotor bias be increased from 3 degrees to 5 degrees which would result in a maximum TRIMPP of 20.5 degrees.

Give a Little, Take a Little

Helicopter design and performance are a series of compromises. In mid-2014, the decision was made to increase the maximum TRIMPP from 18.5 degrees to 20.5 degrees. The 20.5 degree TRIMPP was intended to increase aircraft performance at high density altitudes and high gross weights; that is, deliver more payload at higher altitudes. However, the tail rotor paddles have a fixed pitch range of approximately 31 to 32 degrees due to the throw of the pitch change shaft. If 2 degrees of pitch is added to the maximum TRIMPP, then the minimum TRIMPP (full down collective and full right yaw pedal) is reduced by 2 degrees. The effect of losing 2 degrees of the minimum tail rotor impressed pitch is only observed during light gross weight, low power maneuvers such as autorotations, taxi turns, and other maneuvers with increased right pedal requirements such as sideward flight to the left. To determine if sufficient right directional control could be maintained during low gross weight, low power maneuvers, the test aircraft was stripped of all unnecessary equipment to get the aircraft as light as possible to include reducing the takeoff fuel load. The optimized tail rotor rig procedure was used to set 20.5 degrees maximum TRIMPP on all 4 tail rotor blades. In late 2014, a series of low power maneuvers were flown at light gross weight to verify sufficient right directional control could be maintained. Results of that testing indicated that balanced flight (aircraft trim ball centered) could not be maintained on all the maneuvers but aircraft control was never in question.

In parallel with the optimized tail rotor rig procedure and the 20.5 degrees maximum TRIMPP testing, the test team was planning to conduct testing for flight operations at 105 percent N_R. Though the 105 percent N_R testing was being conducted to increase tail rotor performance at high density altitude and high gross weight, testing would also involve a handling qualities evaluation, an engine-airframe compatibility evaluation, flight performance testing, and a flight loads survey. Most of the flight testing for 105 percent N_R occurred at the Redstone Test Center at Redstone Arsenal, AL. However, to test the high density altitude, high gross weight performance of the tail rotor when operating at 105 percent would require returning to Colorado for a fifth time. For those counting, there was an additional high altitude handling qualities evaluation in Buena Vista, CO in 2013 which was due to an aircraft configuration change and is not discussed in this paper.

Back to Colorado

The fifth trip to Buena Vista, CO occurred during July to August 2015. The objective of the directional control margin testing remained the same as the previous tests flown at Buena Vista in 2012 except directional control margin testing would be flown at 105 percent N_R with 20.5 degrees maximum TRIMPP. The test team was to determine the maximum gross weight the aircraft could be flown while maintaining a 10 percent directional control margin on all 8 relative wind azimuths out to 10 KTAS for both in-ground-effect and out-of-ground-effect flight conditions. The same methodology used in 2012 for determining the initial aircraft test gross weight was applied again in 2015. The initial calculated test referred gross weight was within a couple hundred pounds of the final test referred gross weight. Even though the rotor speed was increased to 105 percent and the maximum TRIMPP was increased to 20.5 degrees, the coefficient of power vs. TRIMPP relationship proved useful in determining the initial test gross weight. During testing of the critical azimuths which were flown first to check if sufficient directional control margin remained, the pilots noted the aircraft was operating at or near the continuous torque limit and operating in the 30 minute turbine gas temperature limit. Because the aircraft was operating in the 30 minute turbine gas temperature limit, the maximum test gross weight was reduced so a power margin was available. As a result of the test gross weight being reduced, there were no directional control margins less than 10 percent on the 8 relative wind azimuths and speeds tested. Results from this test indicated a performance increase of approximately 1,900 pounds referred gross weight when operating at 105 percent N_R with a 20.5 degree maximum TRIMPP when compared to operating at 100 percent N_R with a maximum tail rotor impressed pitch of 18.7 degrees.

CONCLUSION

Black Hawk pilots have historically operated an aircraft that was engine or transmission limited, not tail rotor limited. With proper performance planning, a Black Hawk pilot would not fly into a flight regime where there was insufficient tail rotor authority because engine performance would limit operations in those conditions. With the introduction of the MH-60M with the YT706-GE-700 engine, engine performance permitted the Black Hawk pilot to operate in flight regimes where there was insufficient tail rotor authority. The flight testing conducted on the MH-60M over the last 8 years to address the directional control margin issue have brought the MH-60M full circle. With operations at 105 percent rotor speed and 20.5 degrees maximum TRIMPP, the MH-60M once again flies like any other Black Hawk; except, the MH-60M flies higher and heavier. Though the goal of the directional control margin testing was always to maximize tail rotor performance, reducing the loss of directional control envelope was an added benefit to the warfighter for whom flight test always supports. The test techniques used during MH-60M low airspeed, high altitude testing may prove beneficial to other test teams who wish to maximum helicopter tail rotor performance in the low airspeed, high altitude environment.

TAIL ROTOR DEVELOPMENT CONTINUES

In 2016, the Technology Applications Program Office entered into a Cooperative Research and Development Agreement (CRADA) with Sikorsky Aircraft to develop and flight test a new "wide-chord" tail rotor for the Black Hawk helicopter. The new tail rotor will remain the same diameter as the legacy tail rotor paddles but the chord has been increased by approximately 2.5 inches and incorporates a modern airfoil design (based on the S-76D airfoil). The new tail rotor paddles are currently being manufactured and tested at Sikorsky. Flight testing is scheduled to begin in mid-2017.

Authors:
Jeffrey Von Hor, jeffrey.r.vonhor.civ@mail.mil
Allan Guess, allan.p.guess.civ@mail.mil

ACKNOWLEDGEMENTS

The authors wish to thank the Technology Applications Program Office for their funding and support during the MH-60M flight test program and for their permission to present this paper. Additionally, the authors wish to recognize the many contributions to the MH-60M flight test program from the personnel of the Redstone Test Center, Redstone Arsenal, AL; Lockheed Martin Corporation, Special Operations Forces Contract Logistics Support Services, Lexington, KY; the Special Operations Forces Support Activity, Lexington, KY; and the US Army Research, Development and Engineering Command, Redstone Arsenal, AL.

REFERENCES

[1]Leoni, Ray, "S 70 (YUH-60A UTTAS)", Newsletter, Sikorsky Product History, May 2011, updated 21 June 2012, http://www.sikorskyarchives.com/S-70 (YUH-60A UTTAS).PHP.

[2]Technical Manual, TM 1-1520-280-10, *Operator's Manual for Helicopters, Utility, Tactical Transport, UH-60M, HH-60M*, 14 August 2009, with Change 8, 30 April 2016, p. 5-10.

[3]Technical Manual, TM 1-1520-282-10, *Operator's Manual for MH-60M Helicopter*, 17 September 2016, p. 5-11.

[4]Leoni, Ray, "S 70A: U.S. Army UH-60L Black Hawk, UH-60Q DUSTOFF, MH-6, U.S. Air Force MH-60G/HH-60G Pave Hawk", Newsletter, Sikorsky Product History, June 2011, updated 21 June 2012, http://www.sikorskyarchives.com/S70 L Q K G.php.

[5]Strada, D, "An Investigation of the Effects of Relative Winds Over the Deck on the MH-60S Helicopter During Shipboard Launch and Recovery Operations", The University of Tennessee, Knoxville, May 2002, p. 6.

[6]Leoni, Ray, "S-70A (UH-60M Black Hawk, HH-60M)", Newsletter, Sikorsky Product History, November 2011, updated 25 March 2012, http://www.sikorskyarchives.com/S-70A (UH-60M Black Hawk, HH-60M).php.

[7]Specification, "YT706-GE-700 (CT7-8B5/SOF), Engine Model Specification E2309", 12 May 2005, rev. 18 December 2006, p. 163.

[8]Technical Manual, TM 1-1520-282-10, *Operator's Manual for MH-60M Helicopter*, 17 September 2016, p. 2-53.

Helicopter Turbulence Modeling for Flight Simulation and Handling-Quality Analysis

Renliang Chen
Professor

Honglei Ji
PhD. Candidate

Nanjing University of Aeronautics and Astronautics,
Nanjing, Jiangsu, People's Republic of China

ABSTRACT

This paper presents a distributed turbulence model with accurate spatial correlations for helicopter flight simulation and handling-quality analysis. Firstly, digital filters with longitudinal correlations of the von Karman turbulence are developed to generate discrete turbulence velocity components. Meanwhile, turbulence transverse correlations are considered by relating the filters in different positions with spatial correlations of the von Karman theory. Then, the distributions of both the related filters on the transverse plane in front of helicopter and their velocity components in the longitudinal direction of airspeed, as well as turbulence models of helicopter aerodynamic surfaces are established. Finally, a flight dynamics model coupled with the turbulence model is developed and validated against the flight test data. The effect of turbulence transverse correlations on helicopter handling qualities is analyzed. The results show that neglect of turbulence transverse correlations would aggravate the "rotor-to-body attenuation" effect of main rotor, which results in the underestimate of helicopter roll, pitch and heave rate responses to atmospheric turbulence in the interested frequency range of handling qualities but there is nearly no effect on yaw rate response.

NOTATION

$\boldsymbol{A}, a_{i,j}$	lower triangular matrix and its element in the ith line and jth column
e, R_T	main rotor hinge offset and radius, m
$e^{-\tau s}$	pure time delay of τ seconds
f_i, g_i	coefficients of the difference equations for the longitudinal and transverse turbulence velocity components respectively
H_T, L_T	height and length of turbulence control volume, m
$H_{u,v,w}(s)$	shaping filters of turbulence velocity components to random inputs
i,j,k,l,l_0	integer indices
K, Γ	modified Bessel function of second kind and gamma function
K_P	gain of pilot model
l_g, m_f, n_f	number of grid points of turbulence control volume in the longitudinal, lateral, and vertical directions respectively
$L_{u,v,w}$	scale lengths for turbulence spectra, m
N_b, N_s	number of main rotor blades and number of blade segments
p,q,r	angular rates about body axes, deg/s
r_j	radial station of blade element j, m
$\boldsymbol{R}_{WW}$	correlation coefficient matrix of vertical turbulence components on the transverse plane.
$R_{WW}(i,j)$	element in the ith line and jth column of the matrix $\boldsymbol{R}_{WW}$.
$S(w)$	power spectral density of helicopter response to turbulence
t	time, s
T_L, T_D	time constants in the lead/lag term of the pilot model, s
T_N	neuromuscular system time constant of the pilot model, s
u,v,w	velocity components along body axes, m/s
$\boldsymbol{u}, \boldsymbol{x}$	aircraft control and state vectors
$U_{j,k}(N)$, $V_{j,k}(N)$, $W_{j,k}(N)$	tables storing discrete turbulence components of the filter in the ith line and jth column, N is size of the tables
V	aircraft flight speed, m/s
$V_{\min}, V_{tb}$	minimum and input airspeeds for turbulence filters, m/s
v_0, v_c, v_s	dimensionless main rotor induced velocity components
v_t	dimensionless tail rotor induced velocity
v_x, v_y	dimensionless delayed fuselage downwash and sidewash states
ω, Ω	angular and spatial frequences, rad/s and rad/m
$\boldsymbol{w}$	Atmospheric turbulence velocity vector of all the helicopter aerodynamic surfaces

---

Presented at the AHS Rotorcraft Handling Qualities Technical Meeting, Huntsville, Alabama, USA, February 22-23, 2017.

x,y,z	coordinates of position vector, m
$Y_C(s), Y_C(s)$	linear models of helicopter and pilot
β	aircraft sideslip angle, rad
β_i, ς_i	flap and lag angles of the *i*th main rotor blade (rotating system), rad
$\delta_{col}, \delta_{lat}$, $\delta_{lon}, \delta_{ped}$	collective, lateral, and longitudinal stick inputs as well as pedal input, %
δ_{ij}	Kronecker delta
$\gamma_{u,v,w}$	$V/L_{u,v,w}$ respectively, 1/s
Δd	interval of the grid points in the longitudinal direction, m
Δd_f	interval of the grid points in both the lateral and vertical directions, m
Δt	computing cycle, s
$\Delta u, \Delta v$, Δw	independent turbulence velocity components, m/s
$\Delta U, \Delta V$, ΔW	dependent turbulence velocity components, m/s
ξ_1, ξ_2, ξ_3	longitudinal, lateral and vertical distances of two points
ζ	$\xi/(aL)$, where $a = 1.339$
η_i	independent Gaussian noise source
π	circumference ratio
σ_{FL}	intensity in regions of 1-10 rad/s of helicopter response to turbulence
$\sigma_{u,v,w}$	standard deviations of turbulence velocity components, m/s
Φ, Θ, Ψ	aircraft Euler angles, deg
$\Phi_{u,v,w}(\omega)$	power spectral densities of turbulence velocity components
ϕ_1, ϕ_2	blade tip dynamic twist angle and its rate, rad and rad/s
ψ	azimuth angle of the reference blade, rad
ψ_i, Ψ_i	azimuth angle of the blade *i* in body and wind axes respectively, rad
Ω_T	main rotor rotational speed, rad/s
$E(\bullet)$	mean value of sampling sequence
$(\bullet)_{i,j}$	quantity associated with the blade *i*, element *j*
$(\bullet)_{wf,ht,vt,tr}$	quantities associated with helicopter fuselage, horizontal stabilator, vertical fin and tail rotor respectively
$\dot{(\bullet)}$	first-order derivative
$\{\Delta w^i\}$, $\{\Delta W^i\}$	independent and dependent vertical turbulence velocity sequences
$\lceil\bullet\rceil, \lfloor\bullet\rfloor$	integer ceiling and floor operators

INTRODUCTION

Helicopters are always operated at low altitude which is well within the atmospheric boundary layer. The atmospheric turbulence induced by local terrain and man-made structures would result in undesired helicopter motions and high pilot workload, sometimes even push the aircraft to its operational limits. Therefore, it is very critical to develop an atmospheric turbulence model for helicopter safety and handling-quality analysis.

A variety of physics-based turbulence modeling methods, from the traditional body-fixed method for fixed-wing turbulence modeling (Ref. 1) to the blade-centered and cyclo-stationary random processes (Refs. 2-4), have been developed to predict helicopter response to atmospheric turbulence. The rotating-frame SORBET (Simulation of Rotor Blade Element Turbulence) model developed by McFarland and Duisenberg (Ref. 5) is more suitable for helicopter handling-quality analysis due to the favorable assessment obtained in flight simulation and its high computational efficient. It only generated turbulence velocity components at two selected points along an onset line which is perpendicular to the vehicle flight path and tangential to the leading edge of the main rotor. Turbulence velocity components were then distributed with a transport delay in the longitudinal direction and with the Gaussian interpolation method in the lateral direction. One deficiency of the model is "in the reduced level of lateral, longitudinal, and yaw turbulence cues" (Ref. 5). Based on the SORBET model, the distributed turbulence model was developed by Ji and Chen (Ref. 6) to form an atmospheric turbulence field over all of the helicopter aerodynamic surfaces. The results of helicopter response to atmospheric turbulence calculated by the distributed model showed a good agreement with the corresponding flight test data. However, both the SORBET and distributed turbulence models didn't provide a precise modeling of turbulence transverse correlations. In addition, both the models were based on the Dryden spectra, while the more accurate von Karman spectra were not used. An alternative means of helicopter turbulence modeling is the flight-based Control Equivalent Turbulence Input (CETI) approach (Refs. 7-9). It did not generate turbulence velocity directly, but rather generated equivalent control inputs, which produced the same effect on the vehicle as turbulence itself. Due to the extraction from flight test data, the model was automatically validated for specific helicopter type, but the CETI model couldn't be used widely for all the flight conditions and different kinds of helicopters.

This paper presents a distributed turbulence model with accurate spatial correlations for helicopter handling-quality analysis. Digital filters with rational spectra based on the longitudinal correlations of the von Karman turbulence were developed to generate turbulence velocity components. Turbulence transverse correlations were then modeled by relating the filters in different positions with the spatial correlations of the von Karman theory. These related filters were placed in front of a helicopter so that their turbulence velocity components could be distributed backward to each of the helicopter aerodynamic surfaces as the helicopter was moving forward. Finally, the effect of turbulence transverse correlations on helicopter handling qualities was analyzed.

DISTRIBUTED TURBULENCE MODEL WITH ACCURATE SPATIAL CORRELATIONS

As shown in Figure 1, the atmospheric turbulence field is represented as a control volume of the cuboid *ABCDEFGH* which covers the whole helicopter and moves together with the helicopter. The front surface *ABCD* sits in front of main rotor and is perpendicular to the flight speed *V*. The breadth of the rectangle is $2R_T$ and the height is H_T, where R_T is the radius of main rotor, and H_T is dependent on the total height of the helicopter. Related filters with spatial correlations of the von Karman theory are placed on the surface *ABCD*. From Taylor's hypothesis (Ref. 5), the turbulent flow is stationary and homogeneous in the longitudinal direction. The turbulence filter outputs are statistically valid at either fixed spatial locations or the points translating with the vehicle. In this study, the turbulence velocity components are generated by the related filters with the Gaussian random inputs and are fixed at their inertial positions in the space where they are generated. When the helicopter and cuboid together are moving forward with the flight speed *V*, the turbulence velocity histories are distributed backward to each of the helicopter aerodynamic surfaces with a transport delay. The lateral and vertical distributions of the turbulence velocity components over the cuboid volume are achieved with an interpolation method. Thus, an atmospheric turbulence field is formed for helicopter flight simulation and handling-quality analysis.

Digital Modeling of Related Filters

Firstly, high-order shaping filters with rational approximant to the von Karman spectra are derived by the least-squares fitting method. The von Karman spectra are defined analytically in MIL-F-8785C with the spatial frequency. For an aircraft flying at the speed *V* through a frozen turbulence field with the spatial frequency Ω rad/m, turbulence spectra functions with the angular frequency ω can be gotten with the transforms $\omega(\mathrm{rad/s}) = \Omega(\mathrm{rad/m}) \cdot V(\mathrm{m/s})$ and $\Phi_{u,v,w}(\omega) = \Phi_{u,v,w}(\Omega)/V$, where Φ_u, Φ_v, Φ_w are the power spectral densities of longitudinal, lateral and vertical turbulence velocity components. After scaled by $\Phi_u(0)$, $\Phi_v(0)$, $\Phi_w(0)$ respectively, the normalized spectra functions have the form,

$$\frac{\Phi_u(\omega)}{\Phi_u(0)} = \frac{1}{\left[1+(1.339\bar{\omega}_u)^2\right]^{5/6}} \tag{1}$$

$$\frac{\Phi_v(\omega)}{\Phi_v(0)} = \frac{1+\frac{8}{3}(1.339\bar{\omega}_v)^2}{\left[1+(1.339\bar{\omega}_v)^2\right]^{11/6}} \tag{2}$$

$$\frac{\Phi_w(\omega)}{\Phi_w(0)} = \frac{1+\frac{8}{3}(1.339\bar{\omega}_w)^2}{\left[1+(1.339\bar{\omega}_w)^2\right]^{11/6}} \tag{3}$$

where $\Phi_u(0) = 2\sigma_u^2/(\pi\gamma_u)$, $\Phi_{v,w}(0) = \sigma_{v,w}^2/(\pi\gamma_{v,w})$. $\bar{\omega}_{u,v,w} = \omega/\gamma_{u,v,w}$ are the normalized angular frequencies with $\gamma_{u,v,w} = V/L_{u,v,w}$. σ_u, σ_v, and σ_w are the standard deviations of the turbulence velocity components, and L_u, L_v, and L_w are the scale lengths for the turbulence spectra.

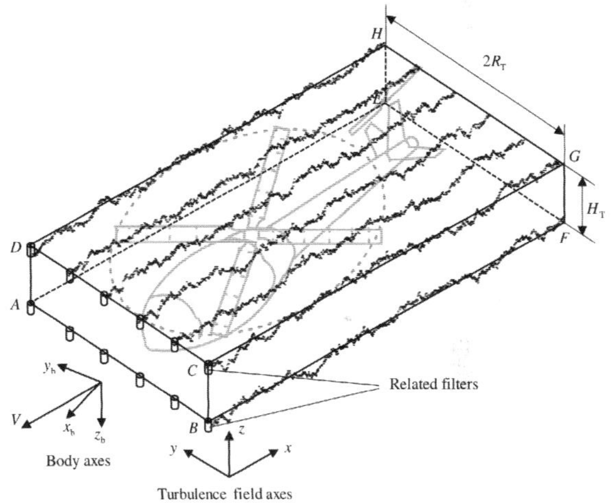

Figure 1. Distributed turbulence field by related filters.

With the formulas (1)-(3), the rational approximant to the theoretical von Karman spectra can be easily gotten by the least-squares fitting method. The resulting high-order shaping filters are

$$H_u(s) = \sigma_u \sqrt{\frac{2}{\pi\gamma_u}} \frac{(1+0.25\gamma_u^{-1}s)(1+0.0244\gamma_u^{-1}s)}{(1+1.19\gamma_u^{-1}s)(1+0.167\gamma_u^{-1}s)(1+0.0170\gamma_u^{-1}s)} \tag{4}$$

$$H_v(s) = \sigma_v \sqrt{\frac{1}{\pi\gamma_v}} \frac{(1+2.618\gamma_v^{-1}s)(1+0.12981\gamma_v^{-1}s)(1+0.0178\gamma_v^{-1}s)}{(1+2.083\gamma_v^{-1}s)(1+0.823\gamma_v^{-1}s)(1+0.08977\gamma_v^{-1}s)(1+0.0129\gamma_v^{-1}s)} \tag{5}$$

$$H_w(s) = \sigma_w \sqrt{\frac{1}{\pi\gamma_w}} \frac{(1+2.618\gamma_w^{-1}s)(1+0.12981\gamma_w^{-1}s)(1+0.0178\gamma_w^{-1}s)}{(1+2.083\gamma_w^{-1}s)(1+0.823\gamma_w^{-1}s)(1+0.08977\gamma_w^{-1}s)(1+0.0129\gamma_w^{-1}s)} \tag{6}$$

A comparison between the longitudinal spectrum of the high-order shaping filters and the Dryden spectrum is shown in Figure 2. The theoretical von Karman spectrum is also included in the figure for comparison. It can be found that the spectrum of the high-order shaping filter agrees well with the theoretical von Karman spectrum in a much wider normalized frequency range than the Dryden spectrum. The comparisons between the lateral and vertical spectra have the same accuracy as the longitudinal spectrum.

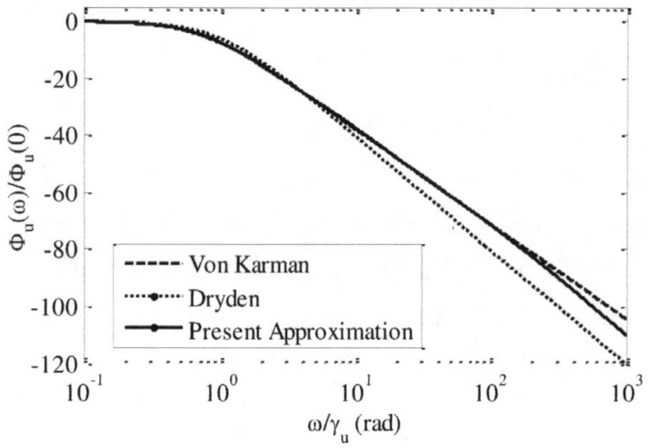

Figure 2. Comparison of longitudinal normalized turbulence spectrum.

To quantitatively analyze the valid frequencies of the filters, a valid normalized frequency range is defined as the frequencies that satisfy the formula

$$\left|10\log_{10}\Phi(\bar{\omega})-10\log_{10}\Phi'(\bar{\omega})\right|\le -3\text{dB} \tag{7}$$

where $\Phi(\bar{\omega})$ represents the theoretical von Karman spectrum, and $\Phi'(\bar{\omega})$ represents the rational approximate spectrum. The valid normalized frequency range measured with Eq. (7) for the Dryden spectra is less than 10 rad. While the valid normalized frequencies are up to 400 rad for the high-order shaping filters.

The digital algorithms to generate independent turbulence velocity components are derived by discretizing the high-order shaping filters in Eqs. (4)-(6) with the Tustin formulation (Ref. 10),

$$H(z)=H(s)\Big|_{s=\frac{2}{\Delta t}\frac{1-z^{-1}}{1+z^{-1}}} \tag{8}$$

where Δt is the computing cycle. The digital algorithms for the discrete turbulence velocity components can be developed by discretizing the turbulence transfer functions with Eq. (8). However, one caution should be mentioned concerning the use of the algorithms that the difference equations presented here were derived by assuming zero initial conditions (Ref. 10). Thus, when the approach is used for the longitudinal velocity component, the resulting difference equations take the following form,

$$\begin{aligned}
\Delta u_1 &= \frac{f_4}{f_0}\eta_1 \\
\Delta u_2 &= \frac{1}{f_0}\left[-f_1\Delta u_1 + f_4\eta_2 + f_5\eta_1\right] \\
\Delta u_3 &= \frac{1}{f_0}\left[-f_1\Delta u_2 - f_2\Delta u_1 + f_4\eta_3 + f_5\eta_2 + f_6\eta_1\right] \\
\Delta u_n &= \frac{1}{f_0}\begin{bmatrix}-f_1\Delta u_{n-1} - f_2\Delta u_{n-2} - f_3\Delta u_{n-3} \\ +f_4\eta_n + f_5\eta_{n-1} + f_6\eta_{n-2} + f_7\eta_{n-3}\end{bmatrix}, \\
& n = 4,5,6,\cdots
\end{aligned} \tag{9}$$

A similar approach can be taken for the transfer function of the lateral turbulence velocity component. The resulting difference equations are

$$\begin{aligned}
\Delta v_1 &= \frac{g_5}{g_0}\eta_1 \\
\Delta v_2 &= \frac{1}{g_0}\left[-g_1\Delta v_1 + g_5\eta_2 + g_6\eta_1\right] \\
\Delta v_3 &= \frac{1}{g_0}\left[-g_1\Delta v_2 - g_2\Delta v_1 + g_5\eta_3 + g_6\eta_2 + g_7\eta_1\right] \\
\Delta v_4 &= \frac{1}{g_0}\begin{bmatrix}-g_1\Delta v_3 - g_2\Delta v_2 - g_3\Delta v_1 \\ +g_5\eta_4 + g_6\eta_3 + g_7\eta_2 + g_8\eta_1\end{bmatrix} \\
\Delta v_n &= \frac{1}{g_0}\begin{bmatrix}-g_1\Delta v_{n-1} - g_2\Delta v_{n-2} - g_3\Delta v_{n-3} - g_4\Delta v_{n-4} \\ +g_5\eta_n + g_6\eta_{n-1} + g_7\eta_{n-2} + g_8\eta_{n-3} + g_9\eta_{n-4}\end{bmatrix}, \\
& n = 5,6,7\cdots
\end{aligned} \tag{10}$$

The difference equations for the vertical turbulence velocity component is similar with that in Eq. (10). The coefficients f_i and g_i of the difference equations for the turbulence velocity components are presented in Tables 1 and 2 respectively.

The total number of the related filters arranged on the plane *ABCD* is assumed to be *N*. *N* sets of independent vertical turbulence velocity sequences with zero mean value and the same standard deviation σ_w can be generated by Eqs. (9) and (10), which are denoted as $\{\Delta w^1\},\{\Delta w^2\},\cdots,\{\Delta w^N\}$.

The vertical turbulence velocity component is taken as an example to show how to relate the independent turbulence velocity component sequences for the formation of related filters. According to the von Karman spectra (Ref. 11), the spatial correlations between the dependent vertical turbulence velocity sequences $\{\Delta W^1\},\{\Delta W^2\},\cdots,\{\Delta W^N\}$ of the *N* related filters on the plane *ABCD* have the form,

$$R_{WW}(i,j)=\frac{2^{2/3}}{\Gamma(1/3)}\zeta^{1/3}\left\{K_{1/3}(\zeta)-K_{2/3}(\zeta)\frac{\xi_2^2+\xi_1^3}{2aL_w\xi}\right\} \quad (11)$$

where $\xi_1=|x_i-x_j|$, $\xi_2=|y_i-y_j|$, $\xi_3=|z_i-z_j|$, $\xi=\sqrt{\xi_1^2+\xi_2^2+\xi_3^2}$, $\zeta=\xi/(aL_w)$, and a = 1.339. x_i, y_i, z_i and x_j, y_j, z_j are the coordinates of the *i*th and *j*th filters on the plane *ABCD*, Γ is the gamma function, and K is the modified Bessel function of second kind.

Table 1. Coefficients of the difference equations for longitudinal turbulence.

$$\gamma_u=\frac{V}{L_u},\alpha=\sigma_u\sqrt{\frac{2}{\Delta t\gamma_u}},\beta=\frac{\Delta t\gamma_u}{2}$$

$$a_1=0.25,a_2=0.0244,b_1=1.19,b_2=0.167,b_3=0.0170$$

$$f_0=(\beta+b_1)(\beta+b_2)(\beta+b_3)$$

$$f_1=(\beta+b_1)(\beta+b_2)(\beta-b_3)+(\beta-b_1)(\beta+b_2)(\beta+b_3)+(\beta+b_1)(\beta-b_2)(\beta+b_3)$$

$$f_2=(\beta+b_1)(\beta-b_2)(\beta-b_3)+(\beta-b_1)(\beta+b_2)(\beta-b_3)+(\beta-b_1)(\beta-b_2)(\beta+b_3)$$

$$f_3=(\beta-b_1)(\beta-b_2)(\beta-b_3)$$

$$f_4=\alpha\beta(\beta+a_1)(\beta+a_2)$$

$$f_5=\alpha\beta\left[(\beta+a_1)(\beta+a_2)+(\beta-a_1)(\beta+a_2)+(\beta+a_1)(\beta-a_2)\right]$$

$$f_6=\alpha\beta\left[(\beta+a_1)(\beta-a_2)+(\beta-a_1)(\beta+a_2)+(\beta-a_1)(\beta-a_2)\right]$$

$$f_7=\alpha\beta(\beta-a_1)(\beta-a_2)$$

Table 2. Coefficients of the difference equations for lateral and vertical turbulence.

$$\gamma_{v,w}=\frac{V}{L_{v,w}},\alpha=\sigma_{v,w}\sqrt{\frac{1}{\Delta t\gamma_{v,w}}},\beta=\frac{\Delta t\gamma_{v,w}}{2}$$

$$a_1=2.618,a_2=0.12981,a_3=0.0178,b_1=2.083,b_2=0.823,b_3=0.08977,b_4=0.0129$$

$$g_0=(\beta+b_1)(\beta+b_2)(\beta+b_3)(\beta+b_4)$$

$$\begin{aligned}g_1=&(\beta-b_1)(\beta+b_2)(\beta+b_3)(\beta+b_4)+(\beta+b_1)(\beta-b_2)(\beta+b_3)(\beta+b_4)\\&+(\beta+b_1)(\beta+b_2)(\beta-b_3)(\beta+b_4)+(\beta+b_1)(\beta+b_2)(\beta+b_3)(\beta-b_4)\end{aligned}$$

$$\begin{aligned}g_2=&(\beta-b_1)(\beta-b_2)(\beta+b_3)(\beta+b_4)+(\beta-b_1)(\beta+b_2)(\beta-b_3)(\beta+b_4)\\&+(\beta-b_1)(\beta+b_2)(\beta+b_3)(\beta-b_4)+(\beta+b_1)(\beta-b_2)(\beta-b_3)(\beta+b_4)\\&+(\beta+b_1)(\beta-b_2)(\beta+b_3)(\beta-b_4)+(\beta+b_1)(\beta+b_2)(\beta-b_3)(\beta-b_4)\end{aligned}$$

$$\begin{aligned}g_3=&(\beta+b_1)(\beta-b_2)(\beta-b_3)(\beta-b_4)+(\beta-b_1)(\beta+b_2)(\beta-b_3)(\beta-b_4)\\&+(\beta-b_1)(\beta-b_2)(\beta+b_3)(\beta-b_4)+(\beta-b_1)(\beta-b_2)(\beta-b_3)(\beta+b_4)\end{aligned}$$

$$g_4=(\beta-b_1)(\beta-b_2)(\beta-b_3)(\beta-b_4)$$

$$g_5=\alpha\beta(\beta+a_1)(\beta+a_2)(\beta+a_3)$$

$$g_6=\alpha\beta\begin{bmatrix}(\beta+a_1)(\beta+a_2)(\beta+a_3)+(\beta-a_1)(\beta+a_2)(\beta+a_3)\\+(\beta+a_1)(\beta-a_2)(\beta+a_3)+(\beta+a_1)(\beta+a_2)(\beta-a_3)\end{bmatrix}$$

$$g_7=\alpha\beta\begin{bmatrix}(\beta+a_1)(\beta-a_2)(\beta-a_3)+(\beta-a_1)(\beta+a_2)(\beta-a_3)\\+(\beta-a_1)(\beta-a_2)(\beta+a_3)+(\beta+a_1)(\beta+a_2)(\beta-a_3)\\+(\beta+a_1)(\beta-a_2)(\beta+a_3)+(\beta+a_1)(\beta-a_2)(\beta-a_3)\end{bmatrix}$$

$$g_8=\alpha\beta\begin{bmatrix}(\beta-a_1)(\beta-a_2)(\beta-a_3)+(\beta+a_1)(\beta-a_2)(\beta-a_3)\\+(\beta-a_1)(\beta+a_2)(\beta-a_3)+(\beta-a_1)(\beta-a_2)(\beta+a_3)\end{bmatrix}$$

$$g_9=\alpha\beta(\beta-a_1)(\beta-a_2)(\beta-a_3)$$

The correlation coefficient matrix $\boldsymbol{R}_{WW} = \left[R_{WW}(i,j)\right]$ between the related filters in different spatial positions can be obtained with Eq. (11). Since the correlation coefficient matrix is positive and symmetric, a lower triangular matrix $\boldsymbol{A}$ can be produced with the Cholesky decomposition (Ref. 12) of the matrix $\boldsymbol{R}_{WW}$, which satisfies the equation

$$\boldsymbol{R}_{WW} = \boldsymbol{A}\boldsymbol{A}^T \tag{12}$$

with

$$\boldsymbol{A} = \begin{bmatrix} a_{11} & 0 & \cdots & 0 \\ a_{21} & a_{22} & \cdots & 0 \\ \vdots & \vdots & \ddots & \vdots \\ a_{N1} & a_{N2} & \cdots & a_{NN} \end{bmatrix}$$

The dependent turbulence velocity sequences $\{\Delta W^1\}$, $\{\Delta W^2\},\cdots,\{\Delta W^N\}$ can be obtained by relating the independent sequences with a transformation as follows,

$$\Delta \boldsymbol{W} = \boldsymbol{A}\Delta \boldsymbol{w} \tag{13}$$

where $\Delta \boldsymbol{W} = \left(\Delta W^1, \Delta W^2, \cdots, \Delta W^N\right)^T$ and $\Delta \boldsymbol{w} = \left(\Delta w^1, \Delta w^2, \cdots, \Delta w^N\right)^{\mathrm{T}}$.

It can be proven that the turbulence velocity sequences $\{\Delta W^1\}$, $\{\Delta W^2\},\cdots,\{\Delta W^N\}$ have a correlation coefficient matrix $\boldsymbol{R}_{WW}$. Since $E\left(\Delta w^i \Delta w^j\right) = \delta_{ij}\sigma_w^2$, where δ_{ij} is the Kronecker delta, the expected values for the dependent variables are given by

$$\begin{aligned} E\left(\Delta W^i \Delta W^j\right) &= E\left\langle \left(\sum_{k=1}^{i} a_{ik}\Delta w_k\right)\cdot\left(\sum_{k=1}^{j} a_{jk}\Delta w_k\right)\right\rangle \\ &= \sum_{k=1}^{\min(i,j)} a_{ik}a_{jk}E\left[\left(\Delta w^k\right)^2\right] = \sigma_w^2 \sum_{k=1}^{\min(i,j)} a_{ik}a_{jk} \end{aligned} \tag{14}$$

With Eq. (12) it can be gotten that

$$\sum_{k=1}^{\min(i,j)} a_{ik}a_{jk} = R_{WW}(i,j) \tag{15}$$

Substitute Eq. (15) in to Eq. (14), we obtain

$$E\left(\Delta W^i \Delta W^j\right) = \sigma_w^2 R_{WW}(i,j) \tag{16}$$

Therefore, the correlation coefficient $R'_{WW}(i,j)$ between the turbulence velocity sequences $\{\Delta W^i\}$ and $\{\Delta W^j\}$ is

$$\begin{aligned} R'_{WW}(i,j) &= \frac{E\left(\Delta W^i \Delta W^j\right)}{\sqrt{E\left[\left(\Delta W^i\right)^2\right]}\cdot\sqrt{E\left[\left(\Delta W^j\right)^2\right]}} \\ &= \frac{R_{WW}(i,j)\sigma_w^2}{\sqrt{R_{WW}(i,i)\sigma_w^2}\cdot\sqrt{R_{WW}(j,j)\sigma_w^2}} = R_{WW}(i,j) \end{aligned} \tag{17}$$

Hence, it is easy to see that the dependent turbulence velocity sequences $\{\Delta W^1\}$, $\{\Delta W^2\}$, $\cdots,\{\Delta W^N\}$ have a correlation coefficient matrix $\boldsymbol{R}_{WW}$.

With the preceding method, the dependent longitudinal and lateral turbulence velocity sequences of the related filters with zero mean values and the same standard deviations σ_u and σ_v can be generated. The related filters with spatial correlations of the von Karman theory are formed.

Filter Distribution on the Transverse Plane

The related filters are distributed to cover the whole plane *ABCD* with the same equal interval in the lateral and vertical directions. In this study, the interval is chosen to make the correlation coefficients between the vertical turbulence velocity components of the adjacent filters larger than 0.95. Based on this principle, the interval Δd_{f} is determined by the scale length for the vertical turbulence spectrum (Ref. 1),

$$\Delta d_{\mathrm{f}} = 0.2L_{\mathrm{w}} \tag{18}$$

With the interval Δd_{f}, the numbers of the filters in each row and column on the plane *ABCD*, m_{f} and n_{f}, are determined,

$$m_{\mathrm{f}} = \left\lceil \frac{2R_{\mathrm{T}}}{\Delta d_{\mathrm{f}}} \right\rceil + 1 \tag{19}$$

$$n_{\mathrm{f}} = \left\lceil \frac{H_{\mathrm{T}}}{\Delta d_{\mathrm{f}}} \right\rceil + 1 \tag{20}$$

where "$\lceil\ \rceil$" denotes the integer ceiling operator.

The vertical turbulence scale length decreases with the decrement of flight altitude (Ref. 1). From Eqs. (19) and (20) it can be gotten that the total numbers of the filters arranged on the plane *ABCD* would increase with the decrement of flight altitude. However, a large number of filters would reduce the computational efficiency. Since the distribution of the aircraft aerodynamic surfaces in the lateral and vertical directions of the turbulence field is not so close, intensive arrangement of the filters is not necessary. The maximum numbers of the filters in each row and column on the plane *ABCD*, m_{fmax} and n_{fmax}, are selected according to the distribution of the aircraft aerodynamic surfaces,

$$m_{\mathrm{f}} = \begin{cases} m_{\mathrm{f}}, & \text{if } m_{\mathrm{f}} \le m_{\mathrm{fmax}} \\ m_{\mathrm{fmax}}, & \text{if } m_{\mathrm{f}} > m_{\mathrm{fmax}} \end{cases} \tag{21}$$

$$n_{\mathrm{f}} = \begin{cases} n_{\mathrm{f}}, & \text{if } n \le n_{\mathrm{fmax}} \\ n_{\mathrm{fmax}}, & \text{if } n > n_{\mathrm{fmax}} \end{cases} \tag{22}$$

Turbulence Distribution in Longitudinal Direction

When the helicopter and cuboid together are moving forward with the airspeed V and computing cycle Δt, the turbulence velocity components are distributed backward to cover each of the helicopter aerodynamic surfaces with a transport delay, as shown in Figure 1. If the turbulence field is divided into a rectangle grid, which has ($l_g \times m_f \times n_f$) grid points with the equal interval $\Delta d = V\Delta t$ in the longitudinal direction of airspeed and Δd_f in the lateral and vertical directions, the discrete turbulence velocity components will always be located at the grid points, and move to next grid points at next computing cycle. Meanwhile the tables $U_{j,k}(l_g)$, $V_{j,k}(l_g)$, and $W_{j,k}(l_g)$ (j = 1, 2, ..., m_f, k = 1, 2, ..., n_f, and l_g is the size of the tables) are created to store the turbulence velocity components generated by the filter in the jth column and kth row on the plane $ABCD$ in reverse order from their last elements to the first circularly. That is, if the turbulence velocity components generated by the filters at the last computing cycle are stored in the ith elements of the circular tables, the newly generated turbulence velocity components will be stored in the (i-1)th elements. When the tables are filled fully as the simulation goes by, the newly generated data of the turbulence velocity components will refresh the tables from their last elements again. There exists a one-to-one mapping relationship between the grid points of the turbulence field in Figure 3 and the turbulence velocity components stored in the tables. At the time $t = n\Delta t$, the latest turbulence velocity components are supposed to be stored in the (l_0+1)th elements of the tables, with

$$l_0 = l_g - n\% l_g \tag{23}$$

where " % " denotes the modulo operator. Then the turbulence velocity components at the grid point (i, j, k) are $U_{j,k}(l)$, $V_{j,k}(l)$, and $W_{j,k}(l)$, with

$$l = \begin{cases} l_0 + i, & if \quad l_0 + i \le l_g \\ l_0 + i - l_g, & if \quad l_0 + i > l_g \end{cases} \tag{24}$$

For a query point which is not located at the grid points, the turbulence velocity components can be regarded to be the same as those of the nearest grid point. Furthermore, the turbulence velocity components generated by the related filters are assumed to be constant over each computing cycle. Therefore, with the coordinates (x, y, z) of the query point in the turbulence field, the turbulence velocity components at the time $t = n\Delta t$ can be determined as $U_{j,k}(l)$, $V_{j,k}(l)$, and $W_{j,k}(l)$, with

$$l = \begin{cases} l_0 + i, & if \quad l_0 + i \le l_g \\ l_0 + i - l_g, & if \quad l_0 + i > l_g \end{cases} \tag{25}$$

$$i = \left\lfloor \frac{x}{\Delta d} \right\rfloor + 1 \tag{26}$$

and

$$j = \left\lfloor \frac{y_{i,j}}{\Delta d_f} + 0.5 \right\rfloor \tag{27}$$

$$k = \left\lfloor \frac{z_{i,j}}{\Delta d_f} + 0.5 \right\rfloor \tag{28}$$

where "$\lfloor \; \rfloor$" denotes the integer floor operator.

Once the table size l_g is established for storing the discrete turbulence velocity components, the minimum airspeed V_{min} for the turbulence filters is determined by distributing the table elements over each of the helicopter aerodynamic surfaces, that is

$$l_g V_{min} \Delta t = L_T \tag{29}$$

where L_T is the total helicopter length. In order to make the tables always cover the whole helicopter, the flight speed V_{tb} for the turbulence filters must be always larger than the minimum airspeed V_{min},

$$V_{tb} = \begin{cases} V & V \ge V_{min} \\ V_{min} & V < V_{min} \end{cases} \tag{30}$$

With the preceding method, the discrete turbulence velocity components generated by the related filters are distributed into the turbulence field.

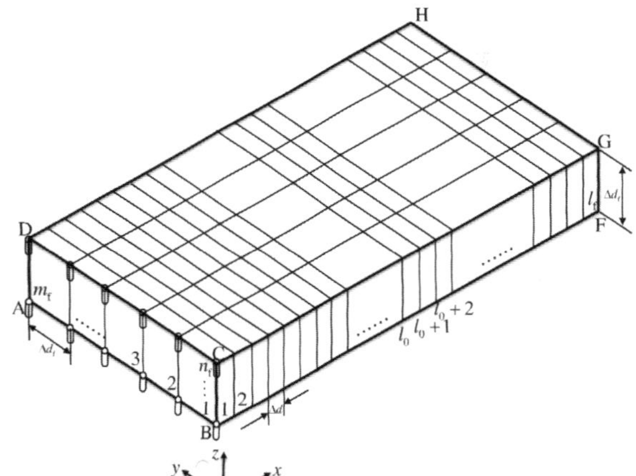

Figure 3. Turbulence field divided by rectangle grid.

Turbulence Models of Helicopter Aerodynamic Surfaces

The model for the turbulence velocity components of each blade element are presented firstly. Due to the rotational velocity of main rotor blades, the position of a blade element in the turbulence field varies with time. For the sake of reducing the amount of computations, the flapping motions of main rotor blades are assumed to be small and the tip path plane of main rotor located on the upper surface *DCGH* of the turbulence field. As shown in Figure 4, the azimuth angle of the blade *i* at the time *t* is given by

$$\psi_i = \int \Omega_T dt + 2\pi\left(\frac{i-1}{N_b}\right) \tag{31}$$

where $2\pi(i-1)/N_b$ is the initial azimuth angle of the blade *i*.

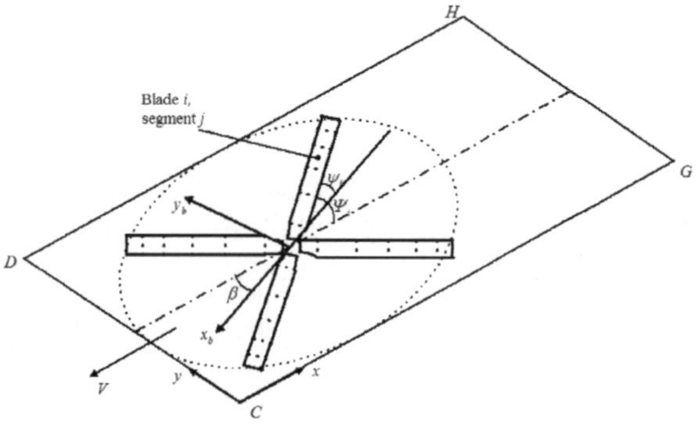

Figure 4. Coordinates of main rotor blade elements.

Helicopter airspeed is not always parallel to the longitudinal body axis, as shown in Figure 4. There is a sideslip angle between them,

$$\beta = \arctan\left(\frac{v}{u}\right) \tag{32}$$

where u and v are the horizontal components of the flight speed V in body axes. Thus the aerodynamic azimuth angle (wind axis) is,

$$\Psi_i = \psi_i + \beta \tag{33}$$

Considering the main rotor hinge offset and lag angle, the coordinates of the blade *i* element *j* on the surface *DCGH* are

$$\begin{aligned} x_{i,j} &= R + e\cos\Psi_i + r_j\cos(\Psi_i + \varsigma_i) \\ y_{i,j} &= R + e\sin\Psi_i + r_j\sin(\Psi_i + \varsigma_i) \end{aligned} \tag{34}$$

where e is the main rotor hinge offset, r_j is the radial station of the blade *i* element *j*, and ς_i is the lag angle of the blade *i*.

With the coordinates of the blade element in the turbulence field, the turbulence velocity components can be obtained with Eqs. (23) and (25)-(28).

The models for the turbulence velocity components of fuselage, horizontal stabilator, vertical fin, and tail rotor are alike. In order to determine the turbulence velocity components at the center pressure of the helicopter aerodynamic surfaces, their coordinates in the turbulence field should be obtained with helicopter geometry parameters and flight states in advance. Then the turbulence velocity components can be obtained with Eqs. (23) and (25)-(28).

SIMULATION MODEL FOR HELICOPTER FLIGHT IN ATMOSPHERIC TURBULENCE

Flight Dynamics Model Coupled with Turbulence Model

A high-order nonlinear helicopter flight dynamics model is used for flight simulation. The detailed formulation of the mathematical model can be obtained in Refs.13-14. The impact of the atmospheric turbulence on helicopter is considered by adding the turbulence velocity components of the helicopter aerodynamic surfaces (including main rotor, fuselage, horizontal stabilator, vertical fin, and tail rotor) directly to the relative incidence velocity components of each surface. For the UH-60 helicopter with a main rotor of four blades, the state-space form of the model is shown as follows

$$\dot{\boldsymbol{x}} = f(\boldsymbol{x}, \boldsymbol{u}, \boldsymbol{w}, t) \tag{35}$$

where $\boldsymbol{x}$ is the aircraft state vector,

$$\boldsymbol{x} = \begin{bmatrix} u\, v\, w\, p\, q\, r\, \Phi\, \Theta\, \Psi\, \dot{\beta}_1\, \dot{\beta}_2\, \dot{\beta}_3\, \dot{\beta}_4\, \beta_1\, \beta_2\, \beta_3\, \beta_4\, \dot{\varsigma}_1\, \dot{\varsigma}_2\, \dot{\varsigma}_3\, \dot{\varsigma}_4 \\ \varsigma_1\, \varsigma_2\, \varsigma_3\, \varsigma_4\, \phi_1\, \phi_2\, v_0\, v_c\, v_s\, v_{tr}\, v_x\, v_y \end{bmatrix}^T \tag{36}$$

u,v,w,p,q,r are the aircraft translational velocity components and angular rates, Φ,Θ,Ψ are the aircraft Euler angles, $\dot{\beta}_1,\dot{\beta}_2,\dot{\beta}_3,\dot{\beta}_4$, $\beta_1,\beta_2,\beta_3,\beta_4$ are the flap angular rates and angles of the four rigid blades, $\dot{\varsigma}_1,\dot{\varsigma}_2,\dot{\varsigma}_3,\dot{\varsigma}_4$, $\varsigma_1,\varsigma_2,\varsigma_3,\varsigma_4$ are the lag angular rates and angles of the four rigid blades, ϕ_1,ϕ_2 are the blade tip dynamic torsion states, v_0,v_c,v_s are the induced velocity components of main rotor, v_{tr} is the induced velocity component of tail rotor, and v_x,v_y are the delayed fuselage downwash and sidewash. $\boldsymbol{u}$ is the control vector,

$$\boldsymbol{u} = \begin{bmatrix} \delta_{col}\ \delta_{lat}\ \delta_{lon}\ \delta_{ped} \end{bmatrix}^T \tag{37}$$

where δ_{col} is the collective stick input, δ_{lat} is the lateral stick input, δ_{lon} is the longitudinal stick input, and δ_{ped} is the pedal input. $\boldsymbol{w}$ is the turbulence velocity vector of the helicopter aerodynamic surfaces,

$$\boldsymbol{w} = \begin{bmatrix} \Delta U_{i,j}\, \Delta V_{i,j}\, \Delta W_{i,j}\, \Delta U_{wf}\, \Delta V_{wf}\, \Delta W_{wf}\, \Delta U_{hs}\, \Delta V_{hs}\, \Delta W_{hs} \\ \Delta U_{vt}\, \Delta V_{vt}\, \Delta W_{vt}\, \Delta U_{tr}\, \Delta V_{tr}\, \Delta W_{tr} \end{bmatrix}^T \tag{38}$$

where $\Delta U_{i,j},\Delta V_{i,j},\Delta W_{i,j}$ are the turbulence velocity components of main rotor blade elements, $\Delta U_{wf},\Delta V_{wf},\Delta W_{wf}$ are the turbulence velocity components of fuselage, $\Delta U_{hs},\Delta V_{hs},\Delta W_{hs}$ are the turbulence velocity components of horizontal stabilator, $\Delta U_{vt},\Delta V_{vt},\Delta W_{vt}$ are the turbulence velocity components of vertical fin, and $\Delta U_{tr}, \Delta V_{tr}, \Delta W_{tr}$ are the turbulence velocity components of tail rotor. The subscripts *i*, *j* indicate that the argument is the turbulence velocity component of the blade *i* element *j*. *t* represents time.

Pilot Model

A simplified precision model of pilot (Ref. 15) is used to control aircraft attitude and trajectory. The pilot model can be described by the transfer function

$$Y_P(s) = K_P \frac{T_L s + 1}{T_D s + 1} \frac{e^{-\tau s}}{T_N s + 1} \tag{39}$$

where K_P is the pilot gain representing the pilot's ability to respond to an error in the magnitude of a controlled variable, T_L is the lead time constant reflecting the pilot's ability to predict a control input and T_D is the lag time constant which describes the ease with which the pilot generates the required input. These three parameters are known collectively as the "human equalization characteristics". The remaining two terms within the transfer function can be defined as the "inherent human limitations" where $e^{-\tau s}$ represents a pure time delay to model pilot's cognitive responsiveness. T_N is the neuromuscular lag time constant which represents the time constant associated with contraction of the muscles through which the control input is applied by the pilot.

VALIDATION OF THE SIMULATION MODEL

Validation of Flight Dynamics Model

Figures 5 and 6 are the comparisons of the calculated roll rate frequency response to lateral control imput with those derived from the flight test data for the UH-60 helicopter in hover (Ref. 16) and forward flight (Ref. 17). For the hover condition, the helicopter is hovering at the altitude of 61 m with the gross weight of 6985 Kg and the 39 degrees of horizontal stabilator incidence angle. For the forward flight condition, the helicopter flies at the altitude of 300 m with the speed of 41 m/s and gross weight of 7153 Kg as well as the zero degrees of horizontal stabilator incidence angle.

From the figures we can find that there is a good agreement between the calculated results and flight test data for the frequencies between 1 and 10 rad/s while there are great errors below 1 rad/s and above 10rad/s. However, they do not affect the application of the model in handling-quality analysis since the interested frequency range for handling-quality investigations is 1-10 rad/s. The on-axis frequency responses of the pitch, yaw, and heave rates have the same accuracy as the roll rate.

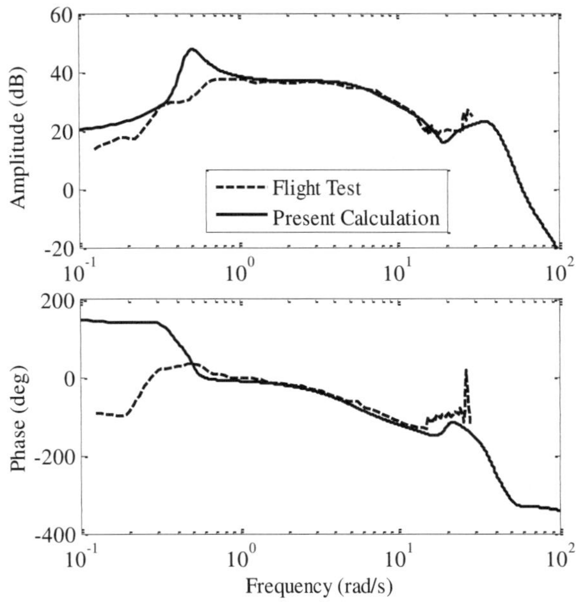

Figure 5. Comparison of roll rate response to lateral control input in hover.

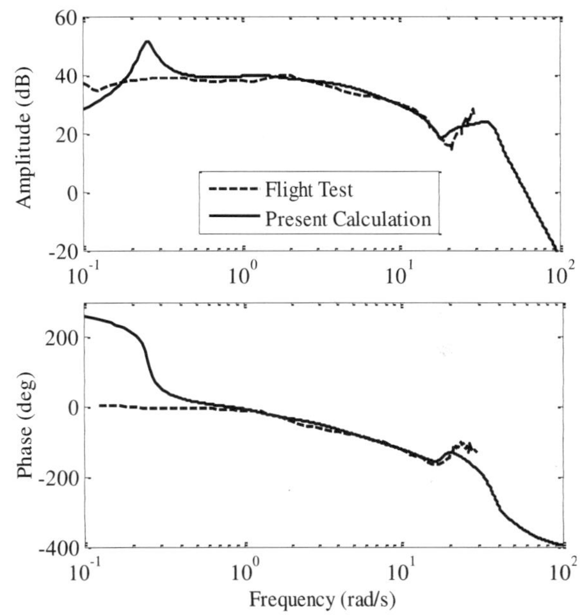

Figure 6. Comparison of roll rate response to lateral control input in forward flight.

Validation of the Related Filters

The spectra and correlation coefficients of the simulated turbulence velocity components by the related filters are calculated and compared to those of the theoretical von Karman turbulence. The flight condition here is the case for a helicopter hovering at the altitude of 12.2 m with the mean wind speed of 11.3 m/s and the vertical turbulence intensity of 1.68 m/s. There are 40 filters arranged in 2 rows on the whole plane ABCD with 20 filters in each row for this flight condition.

Figure 7 shows the comparison between the theoretical and calculated spectra of the vertical turbulence velocity components generated by the related filter 1, filter 10, and filter 20 in the upper row on the plane *ABCD*. Figure 8 presents the comparison between the theoretical and calculated correlation coefficients about the vertical turbulence velocity components of the 20 filters in the upper row. It can be found that all the agreements between the theoretical and simulated spectra are quite satisfactory within the interested frequency range for handling-quality analysis, so is the agreement between the theoretical and simulated correlation coefficients of the related filters. The calculated spectra and correlation cofficients of the longitudinal and lateral turbulence velocity components have the same accuracy.

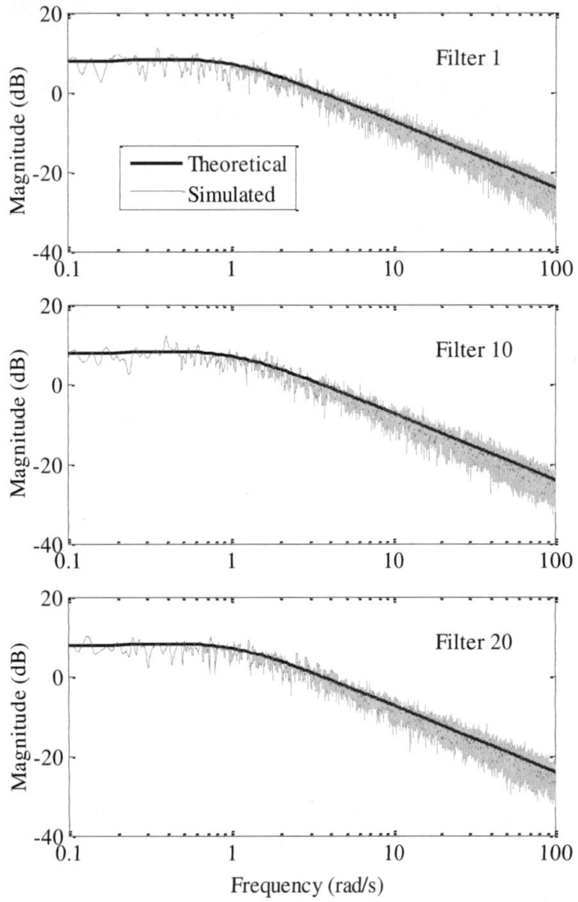

Figure 7. Comparison between theoretical and calculated spectra of vertical turbulence velocity components.

Validation of Helicopter Response to Atmospheric Turbulence

The helicopter responses to turbulence with the Dryden and von Karman spectra are compared and validated against the flight test data in Ref. 8 of the Power Spectral Densities (PSD) for the roll, pitch, yaw, and heave rates. The discrete Dryden filters in Ref. 5 are used to calculate helicopter response to turbulence with the Dryden spectra. The flight condition is the case for a helicopter hovering at the altitude of 12.2 m with the mean wind speed of 11.3 m/s and the vertical turbulence intensity of 1.68 m/s. The results are shown in Figure 9.

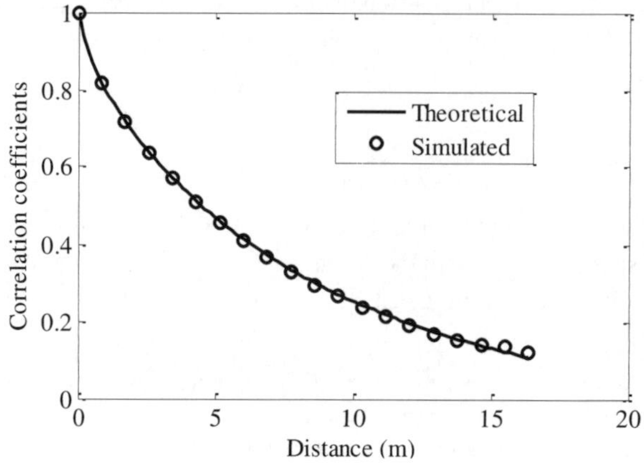

Figure 8. Comparison between theoretical with calculated correlation coefficients of vertical turbulence velocity components.

From the figure it can be seen that the simulated results by the proposed distributed turbulence model with accurate spatial correlations are in good agreement with those from flight test in the interested frequency range of 1-10 rad/s. Therefore, the proposed distributed turbulence model is well suited for flight simulation and the corresponding handling-quality analysis.

Furthermore, it can be seen that for the case in Figure 9, with $\gamma_w = 1.2$, the helicopter response to turbulence with the von Karman spectra is a little larger than that with the Dryden spectra only in the higher frequency range of handling qualities. However for the case in Figure 10, with the flight condition for a helicopter flying at the altitude of 50 m with the speed of 11.3 m/s as well as $\gamma_w = 0.2$, the difference becomes greater and spreads into the lower frequencies. This is because that the valid normalized frequency range of the Dryden filters is only up to 10 rad while those of the proposed high-order shaping filters is up to 400 rad. For the flight conditions with $\gamma_w \geq 1$ both the two kind of filters are valid in the interested frequency range of 1-10 rad/s, while for the latter case with $\gamma_w < 1$ only the proposed high-order shaping filters are valid in the interested frequency range. Note that the turbulence scale lengths increase with flight altitude (Ref. 1), therefore, for the flight conditions with a ratio of flight speed to altitude less than 1, the proposed high-order shaping filters based on the von Karman spectra should be used to replace those with the Dryden spectra in helicopter turbulence model for flight simulation and handling-quality analysis.

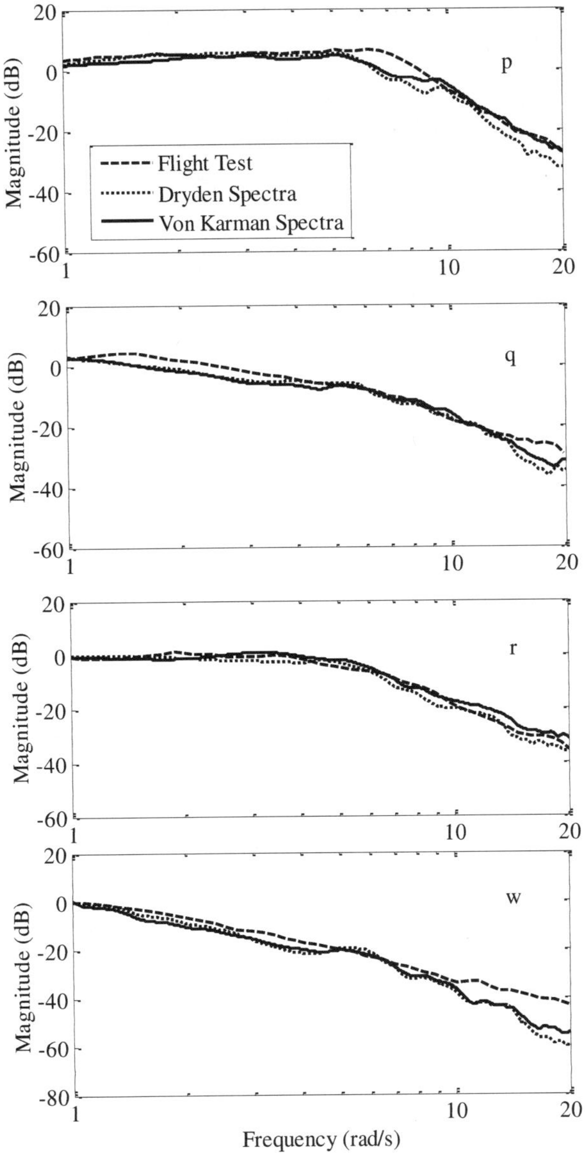

Figure 9. Comparison of helicopter response to turbulence with Dryden and von Karman spectra.

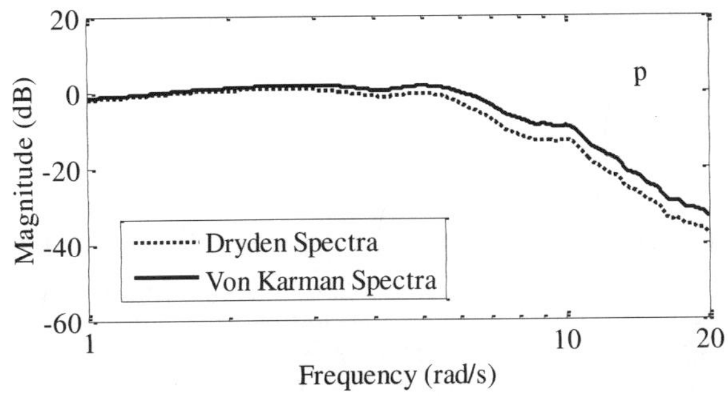

Figure 10. Comparison of roll rate response to turbulence with Dryden and von Karman spectra.

EFFECT OF TURBULENCE TRANSVERSE CORRELATIONS ON HELICOPTER HANDLING-QUALITY ANALYSIS

The effect of turbulence transverse correlations on helicopter handling-quality analysis in atmospheric turbulence is discussed in this section. Smith (Ref. 18) proposed that, in any closed-loop tracking task such as attitude regulation in atmospheric turbulence, rate control activity is of fundamental importance from the standpoint of pilot controlling and perceiving vehicle handling qualities. Based on Smith's fundamental idea, the effect of turbulence transverse correlations on helicopter handling qualities in turbulent atmospheric environment can be analyzed by comparing the helicopter heave and angular rate responses to turbulence with the related and unrelated filters. The unrelated filters are the filters which neglect turbulence transverse correlations and generate independent turbulence velocity components.

Since main rotor provides the primary forces and moments for helicopter flight and control, the turbulence components acting on the main rotor are analyzed firstly. The average vertical turbulence component over the main rotor is estimated in the form

$$\Delta w_{\text{avg}} = \frac{1}{N_{\text{b}} N_{\text{s}}} \sum_{i=1}^{N_{\text{b}}} \sum_{j=1}^{N_{\text{s}}} w_{i,j} \tag{40}$$

where N_{b} is the number of main rotor blades, N_{s} is the number of blade elements. The average roll and pitch turbulence components over the main rotor are estimated with the following equations

$$\Delta p_{\text{avg}} = \frac{180}{\pi} \frac{1}{N_{\text{b}} N_{\text{s}}} \sum_{i=1}^{N_{\text{b}}} \sum_{j=1}^{N_{\text{s}}} \left[\left(y_{i,j} / R_{\text{T}} - 1 \right) \cdot w_{i,j} / 2R_{\text{T}} \right] \tag{41}$$

$$\Delta q_{\text{avg}} = \frac{180}{\pi} \frac{1}{N_{\text{b}} N_{\text{s}}} \sum_{i=1}^{N_{\text{b}}} \sum_{j=1}^{N_{\text{s}}} \left[\left(x_{i,j} / R_{\text{T}} - 1 \right) \cdot w_{i,j} / 2R_{\text{T}} \right] \tag{42}$$

The estimated average vertical, roll and pitch turbulence components acting on the main rotor with the unrelated and related filters are presented in Figure 11 and Figure 12 respectively. From the figures it can be seen that in comparison to the results by the related filters, the average vertical, roll, and pitch turbulence components by the unrelated filters are overly eliminated by the summation of all of the blade element turbulence components. This is because that the lack of turbulence transverse correlations of the unrelated filters increases the randomness of the turbulence components over the main rotor, which aggravates the elimination of the summation of all of the blade element turbulence components.

A PSD analysis of the average vertical, roll and pitch turbulence components acting on the main rotor is conducted and the corresponding results are shown in Figure 13. It can be seen that the difference between the average turbulence components over the main rotor with the unrelated and related filters appears just in the interested frequency range of handling qualities.

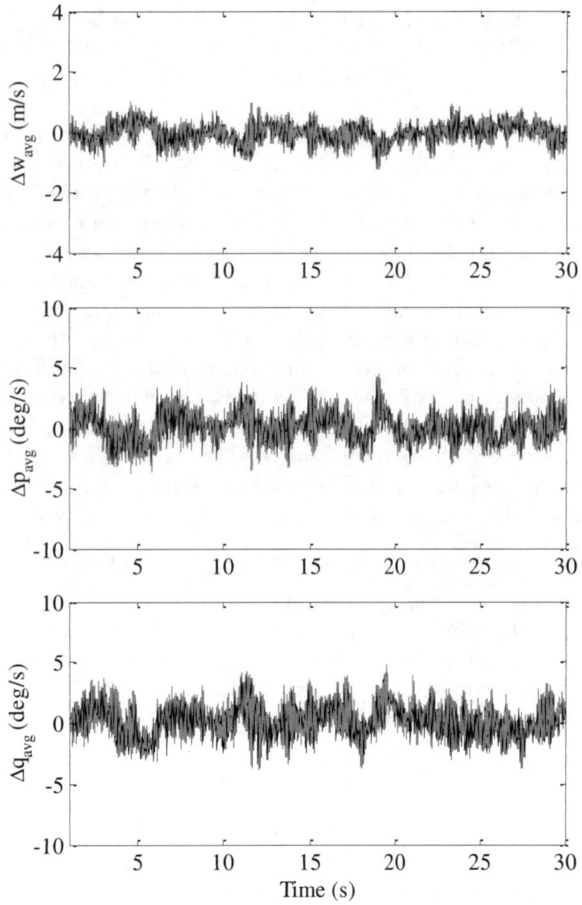

Figure 11. Average turbulence components acting on main rotor with unrelated filters.

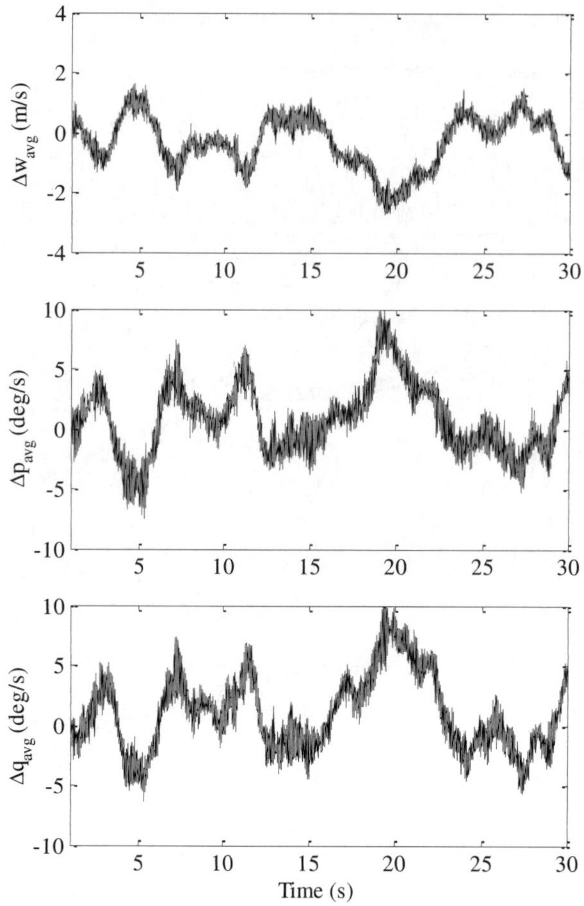

Figure 12. Average turbulence components acting on main rotor with related filters.

The comparison of helicopter roll, pitch, yaw, and heave rate responses to turbulence between the unrelated and related filters are shown in Figure 14. It can be seen that the neglect of turbulence transverse correlations would result in great underestimate of helicopter roll and heave rate responses to atmospheric turbulence in the interested frequency range, and result in a little underestimate of the pitch rate response but there is nearly no effect on the yaw rate response. This is due to the underestimate of the average vertical, roll and pitch turbulence components acting on main rotor with the unrelated filters as shown in Figures 11-13. Main rotor has dominant impacts on both the heave and roll rate responses to turbulence and has the secondary impact on the pitch rate response while has a little impact on the yaw rate response (Ref. 6). This phenomenon reflects the so-called "rotor-to-body attenuation" effect of main rotor (Ref. 5), which indicates that much energy is lost as the turbulence disturbance on main rotor is transferred to body due to the non-uniform turbulence disturbances on different rotor blade elements.

From the preceding analyses, it can be gotten that the neglect of turbulence transverse correlations would aggravate the "rotor-to-body attenuation" effect of main rotor and result in the underestimate of helicopter roll, pitch, and heave rate responses to turbulence. Since turbulence spatial correlations increase with the increment of flight altitude, the effect of turbulence transverse correlations on helicopter handling-quality analysis with flight altitude is further discussed. A ratio between the Root-Mean-Square (RMS) values of the helicopter roll rate response to turbulence with the unrelated and related filters is defined for quantitative analysis,

$$\gamma_p = \frac{\sigma'_p}{\sigma_p} \times 100\% \tag{43}$$

where σ'_p is the RMS value of the roll rate response to turbulence with the unrelated filters, and σ_p is the RMS value of the roll rate response to turbulence with the related filters. From Eq. (43) it can be seen that the effect of turbulence transverse correlations on helicopter roll rate response increases as γ_p decreases.

Figure 15 presents the effect of turbulence transverse correlations on helicopter roll rate response with flight altitude. It can be seen that the neglect of turbulence transverse correlations would result an underestimate of roll rate response to atmospheric turbulence by about 25% for the flight conditions with low flight altitude, and the underestimate increases with the flight altitude and is up to about 40% at the altitude of 300 m. This is because that for the turbulence model with the related filters, the turbulence transverse correlations on the main rotor increase as the flight altitude increases, which results in the reduction of the "rotor-to-body attenuation" effect of main rotor, while for the model with the unrelated filters the turbulence transverse correlations on the main rotor don't vary with flight altitude.

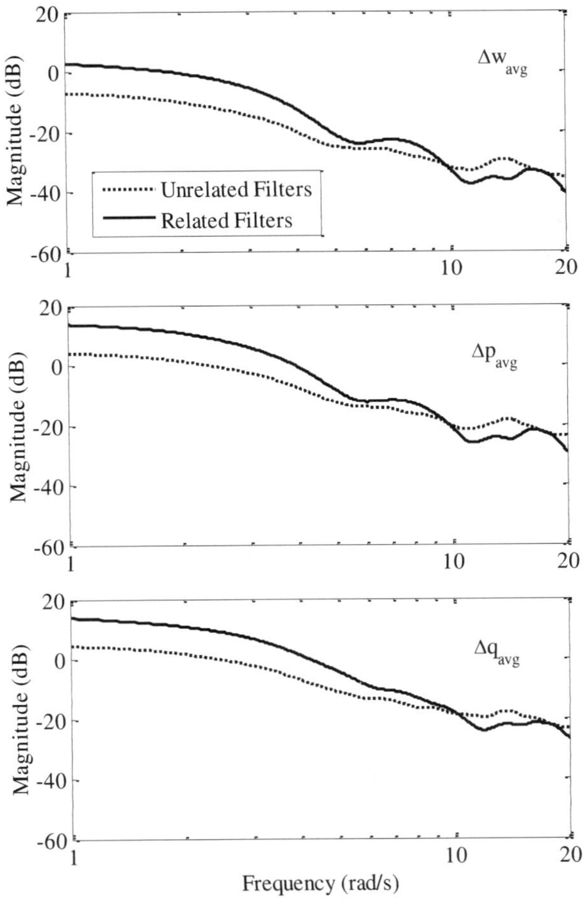

Figure 13. PSDs of average turbulence components acting on main rotor.

CONCLUSIONS

This paper presents a distributed turbulence model with accurate spatial correlations for helicopter flight simulation and handling-quality analysis. Digital filters with longitudinal correlations of the von Karman turbulence are developed. Meanwhile, turbulence transverse correlations are considered by relating the filters in different positions with spatial correlations of the von Karman theory. The effect of turbulence transverse correlations on helicopter handling qualities in atmospheric turbulence is analyzed with the frequency responses of heave and angular rates. The following conclusions can be gotten:

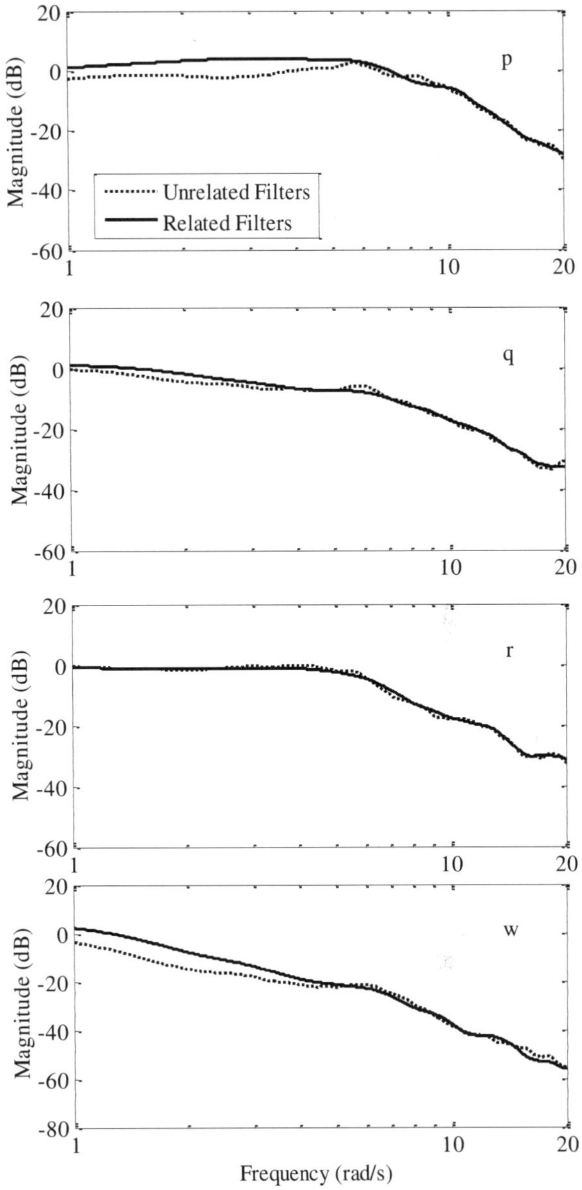

Figure 14. Comparison of helicopter response to turbulence between unrelated and related filters.

1. The spectra and correlation coefficients of the simulated turbulence velocity components show an excellent agreement with those of the theoretical von Karman theory, and a valid normalized frequency up to 400 rad is achieved. The proposed distributed turbulence model based on the related filters can achieve an accurate prediction of helicopter response to atmospheric turbulence in the interested frequency range for handling-quality analysis.

2. Analysis of helicopter handling qualities in atmospheric turbulence by the turbulence models based on the Dryden and von Karman spectra would result in difference results. The difference increases with the increment of flight altitude and the decrement of flight speed. For the flight conditions with a ratio of flight speed to altitude less than 1, the proposed high-order shaping filters based on the von Karman spectra should be used to replace those with the Dryden spectra in helicopter turbulence model for flight simulation and handling-quality analysis.

3. Neglect of turbulence transverse correlations would aggravate the "rotor-to-body attenuation" effect of main rotor, which results in the underestimate of helicopter roll, pitch, and heave rate responses to atmospheric turbulence in the interested frequency range but there is nearly no effect on yaw rate response. For all the flight conditions turbulence transverse correlations have a great impact on helicopter response to atmospheric turbulence, therefore an accurate modeling of turbulence spatial correlations is always important for helicopter handling-quality analysis.

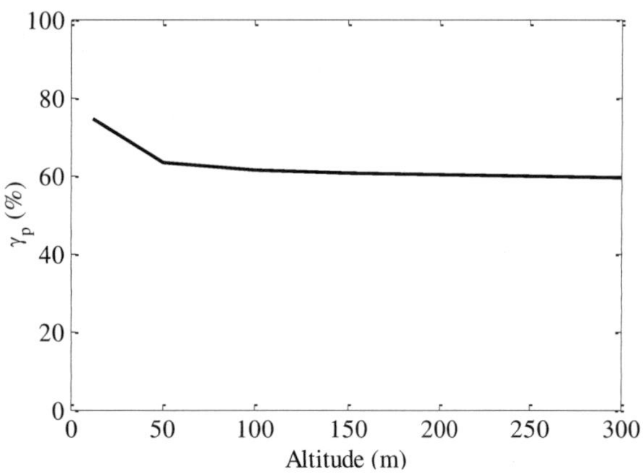

Figure 15. Effect of turbulence transverse correlations on helicopter roll rate response with flight altitude.

ACKNOWLEDGMENTS

This work was supported by the National Natural Science Foundation of China (Grant Number: 11672128).

REFERENCES

[1]"Flying Qualities of Piloted Airplanes," U.S. Department of Defense, MIL-SPEC MIL-F-8785C, Nov. 1980.

[2]Dang, Y. Y., Gaonkar, G. H., Prasad, J. V. R., "Parallel Methods for Turbulence Simulation and Helicopter Response Prediction", Journal of the American Helicopter Society, Vol. 41, No. 3, 1996, pp. 219-231.
doi: 10.4050/jahs.41.219.

[3]van Gool, P. C. A., "Rotorcraft Responses to Atmospheric Turbulence", Ph.D. Dissertation, Technische Unversiteit Delft, Delft, The Netherlands, 1997.

[4]Gaonkar, G. H., "Review of Turbulence Modeling and Related Applications to Some Problems of Helicopter Flight Dynamics", Journal of the American Helicopter Society, Vol. 53, No. 1, 2008, pp. 87-107.
doi: 10.4050/JAHS.53.87.

[5]McFarland, R. E., Duisenberg, K., "Simulation of Rotor Blade Element Turbulence", NASA TM-108862, Washington, D.C., 1995.

[6]Ji, H. L., Chen, R. L., and Li, P., "A Distributed Atmospheric Turbulence Model for Helicopter Flight Simulation and Handling Quality Analysis". Journal of Aircraft, in printed 2016.
doi: 10.2514/1.C033667.

[7]Labows, S. J., "UH-60 Black Hawk Disturbance Rejection Study for Hover/Low Speed Handling Qualities Criteria and Turbulence Modeling", M. S. thesis, Naval postgraduate school, Monterey, CA, 2000.

[8]Lusardi, J. A., Blanken, C. L., and Tischler, M. B., "Piloted Evaluation of a UH-60 Mixer Equivalent Turbulence Simulation Model". Proceedings of 59th Annual Forum of American Helicopter Society, AHS International, Virginia, USA. May 2003, pp. 309-320.

[9]Seher-Weiss, S., Von Gruenhagen, W., "Development of EC 135 Turbulence Models via System Identification", Aerospace Science and Technology, Vol. 23, No. 1, 2012, pp. 43-52.
doi: 10.1016/j.ast.2011.09.008.

[10]Katz, P., "Digital control using microprocessors", Prentice-Hall International, INC, USA. 1981, pp:47-64.

[11]Etkin, B., "Dynamics of Atmospheric Flight", John Wiley and Sons, Inc., New York, USA. 1972, chapter 13.

[12]Ascher, U. M., and Chen, G., "A First Course in Numerical Methods", Society for Industrial and Applied Mathematics, Philadelphia, USA. 2011, pp: 104-107.

[13]Howlett, J. J., "UH-60A Black Hawk Engineering Simulation Program", NASA CR-166309, Washington, D.C., 1981.

[14]Li, P., and Chen, R. L., "A Mathematical Model for Helicopter Comprehensive Analysis", Chinese Journal of Aeronautics, Vol. 23, No. 3, 2010, pp. 320-326.

doi: 10.1016/s1000-9361(09)60222-3.

[15]McRuer, D. T., Jex, H. R., "A Review of Quasi-Linear Pilot Models", IEEE Transactions on Human Factors in Electronics, Vol. 8, No. 3, 1967, pp. 231-249.
doi: 10.1109/THFE.1967.234304.

[16]Ballin, M. G., Dalang-Secrétan, M. A., "Validation of the Dynamic Response of a Blade-Element UH-60 Simulation Model in Hovering Flight", Journal of the American Helicopter Society, Vol. 36, No. 4, 1990, pp. 77-88.
doi: 10.4050/jahs.36.77.

[17]Kim, F. D., Celi, R., Tischler, M. B., "Forward Flight Trim Calculation and Frequency Response Validation of a High-Order Helicopter Simulation Model". Journal of Aircraft, Vol. 30, No. 6, 1993, pp. 854-863.
doi: 10.2514/3.46427.

[18]Smith, R. H., "A Theory for Handling Qualities with Applications to MIL-F-8785B", Air Force Flight Dynamics Laboratory, AFFDL-TR-75-119, Wright-Patterson AFB, OH, Oct. 1976.

AH-64 Apache DVE Mission Workload and Handling Qualities Simulation Study

Gary Klein Senior Engineer

John Krainski Senior Engineer

Dr. Kendra Befort Assoc. Technical Fellow

Dr. Russell Enns Technical Fellow

The Boeing Company
Mesa, AZ

ABSTRACT

Driven by evolving battle tactics to not only operate, but exploit, Degraded Visual Environments (DVE), this paper presents the results of a preliminary study designed to assess the degradation in mission performance when operating under such conditions. Specifically, the impact on pilot workload, situational awareness, pilotage performance and handling qualities when operating in DVE is assessed and compared to operating under Good Visual Environment (GVE) conditions, as well as Visually-Aided DVE (DVE-VA). Results were collected for a challenging, but realistic, attack mission in an AH-64 Apache piloted simulator. The results suggest operational challenges in DVE conditions, limitations of Visually-Aided DVE, and provide evidence of a performance gap that is predicted to be improved with future flight control enhancements. Based on the deficits observed and knowledge of what existing flight control technologies can provide, this paper also makes recommendations for various candidate technologies.

INTRODUCTION

Over the past two decades, Degraded Visual Environment (DVE) has contributed to significant rotorcraft accidents, fatalities and material losses. This has been noted and quantified in a variety of articles and papers, such as references 1, 2 and 3. Dellert summarizes well in reference 3, "... over the past 10 years DVE contributed to 87 rotorcraft accidents, 108 fatalities and more than $880 million in material losses. Many of the DVE losses were due to operations in brownout conditions, which are helicopter-induced dust clouds resulting from downwash of the rotor system. However, DVE is more than just brownout. Other DVE conditions factors include smoke, rain, smog, sand and dust, clouds, darkness, fog, snow and flat light". The loss of situational awareness in DVE is amplified by managing high workload for flight as well as tactical operations.

Furthermore, attack helicopter cockpits have become more complex over time and are expected to continue to evolve to provide additional capabilities to the warfighter, such as battlefield management tasks and further advancements in Manned/Unmanned Teaming (MUM-T). Therefore, the Army's goal, as stated by Major General Michael Lundy, "... We have got to be able to fly and fight no matter what the weather, no matter what the visual conditions are,"[2] sets the bar high for intentional DVE operations and the necessity to improve the current levels of performance and safety in DVE.

The challenge put forth to flight control and crew systems designers is to develop innovative solutions that will reduce workload and enhance situational awareness while continuing to meet aircraft handling qualities requirements for DVE flight. Potential DVE mitigation system solutions are comprised of three pillars: improved flight controls, sensors, and cueing."[5,6]. Therefore, as part of the Apache helicopter modernization effort, Boeing is in the process of defining, developing and evaluating DVE solutions to support the three pillars of DVE that will enable Army Aviation to "Own the Weather".

Initially, an important part of the effort is to develop relevant quantifiable methods that will measure the impacts of various DVE technologies and validate their benefits. Specifically, it is key to define relevant mission tasks under which the effects of pilot workload, handling qualities, situational awareness, pilotage performance and safety in DVE can be measured. Only with such mission tasks can one quantify the adverse effects that operating in DVE has

---

Presented at the AHS Rotorcraft Handling Qualities Technical Meeting, Huntsville, Alabama, USA, February 22-23, 2017.

over operating in Good Visual Environments (GVE) and quantify where there are gaps that need technology solutions.

This paper summarizes the work to develop relevant attack helicopter mission tasks and to assess the impact of operating in DVE compared to operating in GVE. Specifically, this paper describes a test mission, test methodology and assessment metrics defined and developed to measure the impacts of operating in DVE. Next, this paper reports on pilot performance under conditions of GVE, DVE, and Visually-Aided DVE (DVE-VA) to provide baseline performance and illustrate the DVE operational gaps. With these results, recommendations are then made where specific flight control, display and sensor technologies are predicted to benefit pilot workload, handling qualities, situational awareness, and pilotage performance for the various parts of the defined mission tasks.

EXPERIMENT

The purpose of the experiment was to identify the challenges in exploiting DVE for the current AH-64 attack helicopter configuration. This was accomplished by defining relevant attack helicopter mission tasks for measuring pilot workload, situational awareness, pilotage performance, and handling qualities in a challenging, but realistic, attack mission in both GVE and DVE. A secondary objective was to obtain a preliminary assessment of Visually-Aided DVE using a 3-dimensional map on the Helmet Mounted Display (HMD).

This section describes the experiment performed, providing details of the test mission vignettes, the metrics collected, and the simulation environment.

Test Mission Vignettes

The test mission was developed to be operationally relevant, but structured utilizing vignettes and phasing to provide systematic and consistent measurements. In order to impact pilot workload, during the mission, the Co-Pilot/Gunner (CPG) was task-saturated due to UAS management (e.g., MUM-T), and was occasionally required to shed some basic tasks to the pilot, such as entering grid coordinates and radio frequency changes.

The test mission was divided into five vignettes, as detailed in Table 1, with some vignettes consisting of multiple phases. The vignettes were specifically chosen to include elements seen as potentially challenging to pilots while operating in DVE.

The defined test mission starts with Vignette #1 which consists of the pilot preparing for a departure and, while performing a lateral slide, he needs to perform an immediate landing (Phase 1.1) as is sometimes required to conserve fuel while waiting on his wingman or possibly responding to a system malfunction. Once the flight is ready for departure, he executes a hasty lateral reposition to the departure point (Phase 1.2). This vignette is intended to show the difficulty in precisely maneuvering when the visibility to ground reference and/or obstacles is degraded. Phase 1.1 is used to assess the pilot workload associated with having to rapidly arrest lateral motion prior to a vertical landing in order to prevent a roll over or damage to the landing gear. Phase 1.2 was set up to be very similar to the ADS-33 Lateral Reposition MTE so that Cooper Harper Ratings could be taken and compared to other results.

Scenario Vignette	Phase	Assigned Pilot Activity
1. Takeoff/ Emergency Landing, Hasty Lateral Reposition	1.1 Takeoff, Landing	Unplanned Emergency Landing
	1.2 Reposition	N/A
2. Enroute to FARP	2.1 Enroute	Pilot receives and enters updated FARP location and radio frequency, then contacts FARP
3. Approach to FARP with Go-Around, 2nd Approach and Landing	3.1 Approach, Go-Around	Wave-off, go-around at 15 ft AGL
	3.2 2nd Approach, Landing	N/A
4. Takeoff from FARP, Hold and Redirect, Direct to TIC	4.1 Takeoff (Alt/Airspeed)	N/A
	4.2 Hold, Redirect	Pilot receives and enters TIC friendly location and radio frequency
	4.3 Direct to TIC	Pilot sends friendly positions to wingman
5. Wagon Wheel Orbit and Engagement	5.1 Orbit and Engage	CPG chatter to identify and target enemy

Table 1. Test Mission Vignettes

In Vignette #2, the pilot departs and flies along a pre-planned route in mountainous terrain to a Forward Arming and Refueling Point (FARP) with a goal of crossing a passage point at a pre-determined time (Phase 2.1). While flying the 10 minute flight to the FARP, the pilot is required to make radio frequency changes and sends a waypoint update to his wingman. The intent of this vignette was to show the workload comparisons between visual environments in the presence of added mission tasks.

In Vignette #3, on short final for the FARP, the pilot receives instructions for a wave-off and go-around (Phase 3.1) because the landing area is obstructed by a vehicle. The pilot performs a go-around and sets up for another approach to the FARP. He then performs a confined area landing at the FARP (Phase 3.2). While in the FARP, he refuels and rearms up to maximum gross weight. On departure from the FARP he performs an "altitude over airspeed" take off because of the brownout conditions (Phase 4.1). After departure, he continues on the pre-planned route. The main purpose of including this vignette was to illustrate the workload associated with landing in confined, unimproved landing areas in and around ground obstacles such as fuel trucks.

Figure 1. Pilot View with Fog/Mist Limited Visibility.

Vignette #4 consists of the pilot receiving a request for immediate fire support due to Troops In Contact (TIC) and a new destination (Phase 4.2) soon after departure from the FARP. While waiting for further instructions and getting the TIC locations, the pilot enters a holding pattern. Once the friendlies and enemy locations are relayed and entered into the system, the pilot proceeds direct to the TIC as quickly as possible (Phase 4.3). During the transit to the TIC, the pilot sets up an orbit pattern on his Tactical Situation Display (TSD) around the target location. Vignette #4 illustrates the affect DVE has on the ability of the pilot to keep the aircraft within limits and the affect DVE has on performance as well as pilot workload.

Figure 2. Pilot View of Brownout Condition.

Vignette #5 consists of the pilot entering into and establishing an orbit pattern as required to provide fire support for the TIC (Phase 5.1). Vignette #5 was included in the experiment to illustrate the workload associated with maintaining a constant speed altitude orbit pattern while busy with other mission tasks.

This test mission was executed under three visual conditions: GVE, DVE, and Visually-Aided DVE. GVE had unlimited visibility throughout the entire mission. For the purposes of this experiment, DVE consisted of 1000 meter visibility in all directions as well as simulated brownout during takeoffs and landings. Examples of the simulator's DVE conditions due to fog/mist and brownout are shown in Figure 1 and Figure 2. Visually-Aided DVE had the same conditions as DVE while providing the pilot with access to a development version of a 3-D map on the Helmet Mounted Display (HMD) as shown in Figure 3.

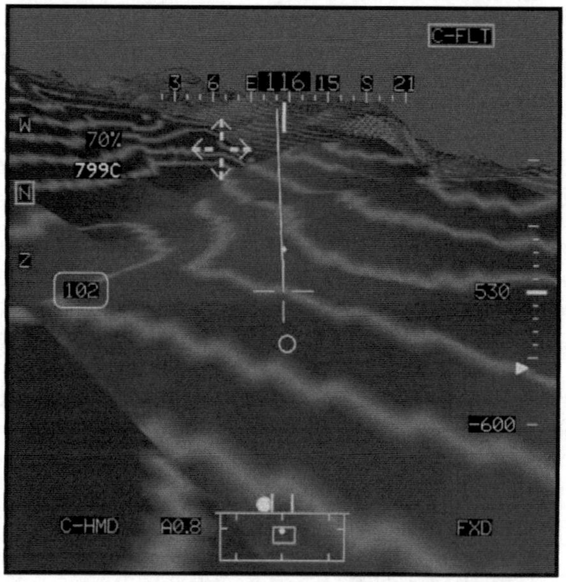

Figure 3. Typical 3-D Map HMD Visual Aiding

Methodology and Metrics

A within-subject design method where all pilots performed in all three environmental conditions was used to hold individual differences consistent across conditions. In addition, counter balancing was used to insure that each pilot started and ended with exposure to a different condition to offset any issues with practice effects or inflated fatigue ratings. The following was collected for analysis:

- Observation Data: Collected during pilot training, mission execution, and pilot debrief
- Scale Data: Completed by pilots after the end of each test vignette
- Performance Data: Recorded and analyzed in post-processing

Data was collected and analyzed from three Boeing test pilots and one former U.S. Army AH-64 pilot now working for Boeing as a subject matter expert. All the pilots have significant flight experience in one or more Apache variants.

The metrics included were pilot workload, situational awareness, handling qualities, and pilotage performance. Pilot workload was assessed using the Bedford workload scale for all vignettes (see Figure 4) [4]. Situational awareness (SA) ratings, using an internally-developed three point Likert-type scale[5], were collected to assess system and tactical situational awareness. Tactical SA was defined as the pilot's global awareness (i.e., common operating picture) and System SA was defined as awareness of the aircraft systems (i.e., states, modes, status, etc.). For the purpose of comparing competing workload metrics, some vignettes included a NASA Task Load Index (TLX) workload assessment (see Figure 5)[6]. Handling Qualities were assessed using Cooper-Harper rating scale[7], but the assessment was limited to the ADS-33E Lateral Reposition maneuver in Vignette 1.2 (see Figure 6). Pilotage performance was evaluated in key areas such as: ability to stay on course, time-to-complete mission elements, and aircraft power management.

After each of the five vignettes, the simulation was paused for the pilots to complete the appropriate relevant metric assessments for workload and situational awareness. At the end of the mission, pilots were debriefed and they completed a brief survey.

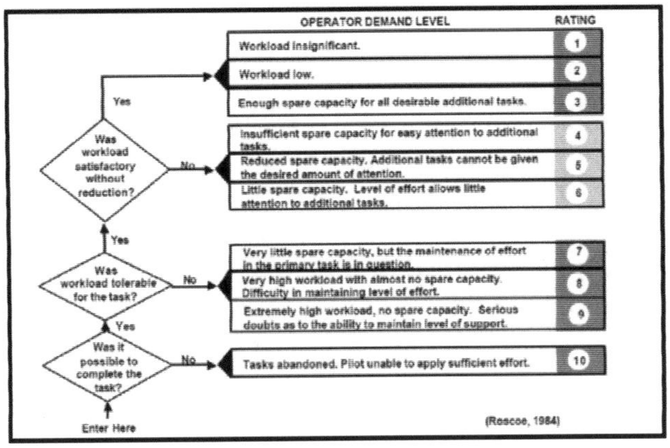

Figure 4. Bedford Workload Rating Scale

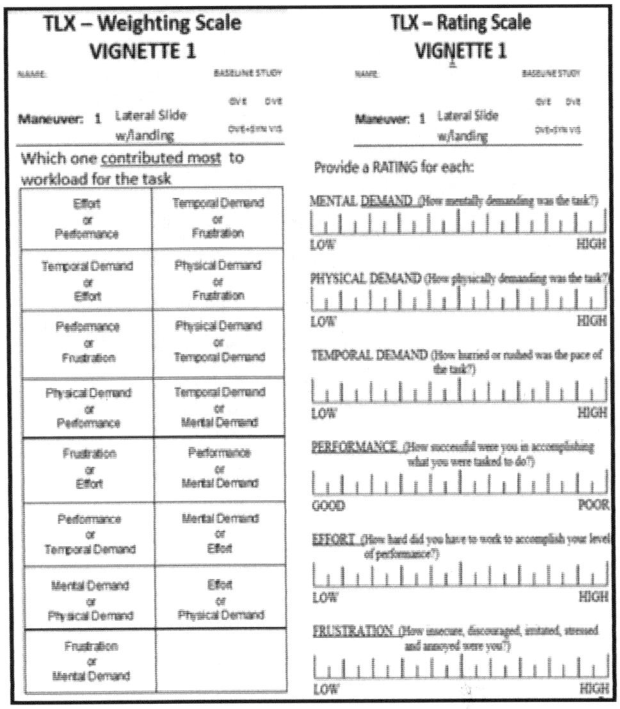

Figure 5. NASA TLX

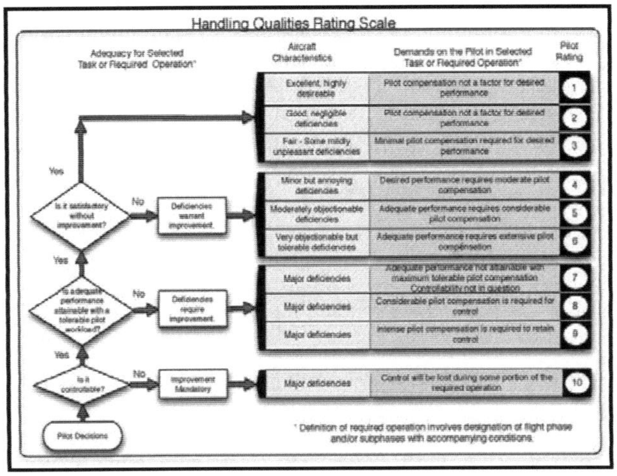

Figure 6. Cooper-Harper HQR Scale

Simulation Environment

The evaluation was performed in the Engineering Development Simulator (EDS) at The Boeing Company in Mesa, Arizona. The EDS is an immersive domed simulator with an Apache cockpit containing actual aircraft hardware and software. Figure 7 shows a view of the EDS pilot cockpit.

Figure 7. EDS Apache Pilot Cockpit

RESULTS

The following describes some of the key results for each of the five vignettes, followed by a summary of the results across all vignettes.

Vignette #1 Results

Vignette #1 was a lateral hovering repositioning and landing intended to show the difficulty in precisely maneuvering when the visibility to ground reference and/or obstacles is degraded. Phase 1.1 specifically measured the workload associated with having to rapidly arrest lateral motion prior to a vertical landing in order to prevent a roll over or damage to the landing gear. For this maneuver, the average workload (TLX and Bedford Scale) for DVE was 60% higher than for GVE. The time to complete the landing was also 57% (4.5 seconds) longer in DVE than GVE. This maneuver is an example where the Landing Descent Rate Limit cueing feature developed and demonstrated in the Adaptive Vehicle Management System (AVMS) program showed significant (27%) workload improvement over the baseline system under GVE conditions and is expected to play even a more significant role in future DVE operations.[8,9] Furthermore, the axes decoupling that an enhanced flight control system provides would also go a long way to addressing this issue.

Phase 1.2 was set up to be very similar to the ADS-33 Lateral Reposition MTE so that Cooper Harper Ratings could be taken and compared to other results. This maneuver resulted in average HQR 1.8 degradation from GVE to DVE. GVE was Level I but DVE was middle to high Level II. As shown in Figure 8a - Figure 8c, the pilot's repositioning route tracking performance worsened as visibility degraded, which was also reflected in the trending higher workload and lower situational awareness data with DVE conditions receiving the highest workload and lowest situational awareness ratings (Figure 12 - Figure 14).

Figures 8a-c show the ADS-33 desired and adequate boundaries for the lateral reposition maneuver, which are the same for operating in both GVE and DVE. It is important to note that in GVE, the position, altitude and heading remained in the "Desired" range over the entire maneuver; however, in DVE, the Desired Boundary was violated for each of the 3 criteria. Furthermore, even the Adequate Boundary was violated for DVE and DVE-VA for two of the three criteria.

Another significant result was the average time to complete the maneuver was 44% (18 seconds) longer in DVE than in GVE due to pilot workload. This is an example of how DVE affects mission performance as well as safe navigation. This maneuver was also one where an enhanced flight control system such as that demonstrated in AVMS can significantly decrease workload and improve course tracking and heading performance.[1,2]

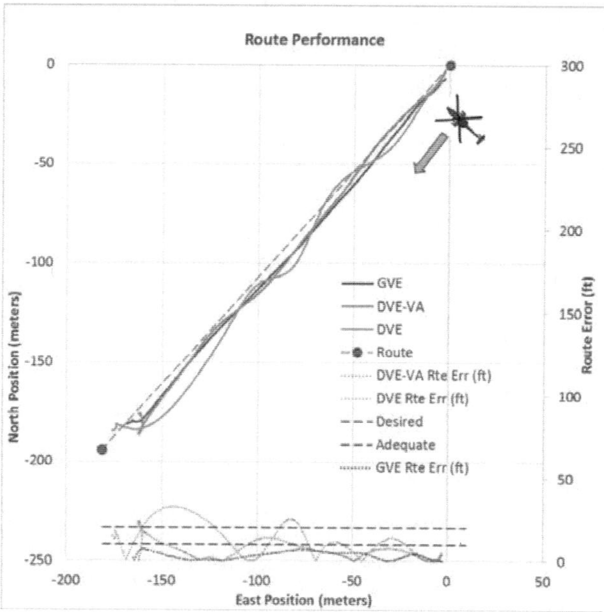

Figure 8a. ADS-33E Lateral Reposition Route Tracking Performance

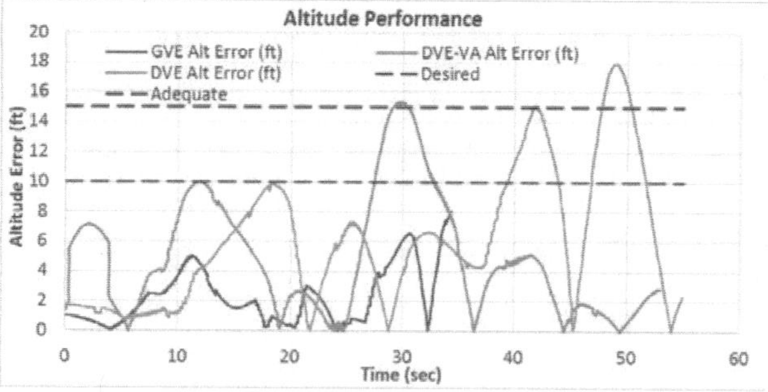

Figure 8b. ADS-33E Lateral Reposition Altitude Performance.

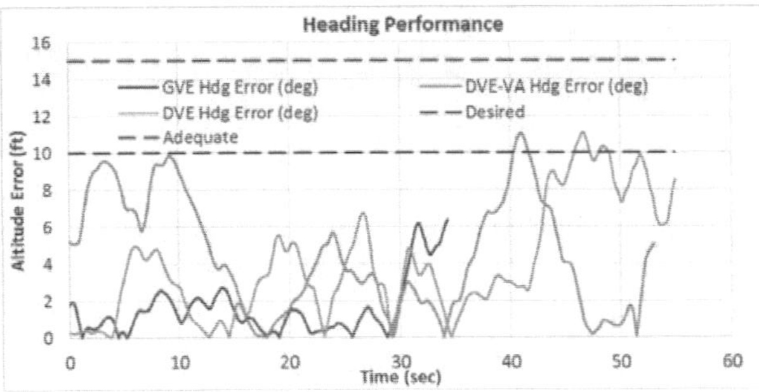

Figure 8c. ADS-33E Lateral Reposition Heading Response.

Vignette #2 Results

Vignette #2 of the mission simulation entailed flying to the FARP, and the intent of evaluating this maneuver was to show the workload comparisons between visual environments in the presence of added mission tasks. Figure 9 clearly shows that adding routine tasks such as updating FARP coordinates and changing radio frequencies to the flying crewmember, has more impact on the pilot's tracking performance the more degraded the visual environment was, which was corroborated by higher workload and reduced situational awareness ratings (Figure 12-Figure 14).

Enroute to the FARP, the DVE increased the average route tracking error by 100% for when visually aided and 143% when not aided compared to GVE. Altitude errors showed similar results. DVE increased pilot workload by 35% when visually aided and by 52% {without aiding} over the GVE case using the Bedford scale. The time to complete tasks such as updating the FARP grid coordinates and contacting the FARP were increased as well (~100% longer for both tasks), which equated to well over a minute of lost time available to do other tasks when in DVE.

The pilots noted that during this vignette, an automated or pilot-by-exception waypoint navigation capability would be valuable to reduce workload while the pilot is performing non-flying tasks. The Pilot-by-Exception concept was demonstrated on AVMS, where the hands off controls (trim) position is automatically moved for various reasons such as:

1. satisfying waypoint navigation routing,
2. avoiding obstacles, and
3. station keeping for formation flying.[8]

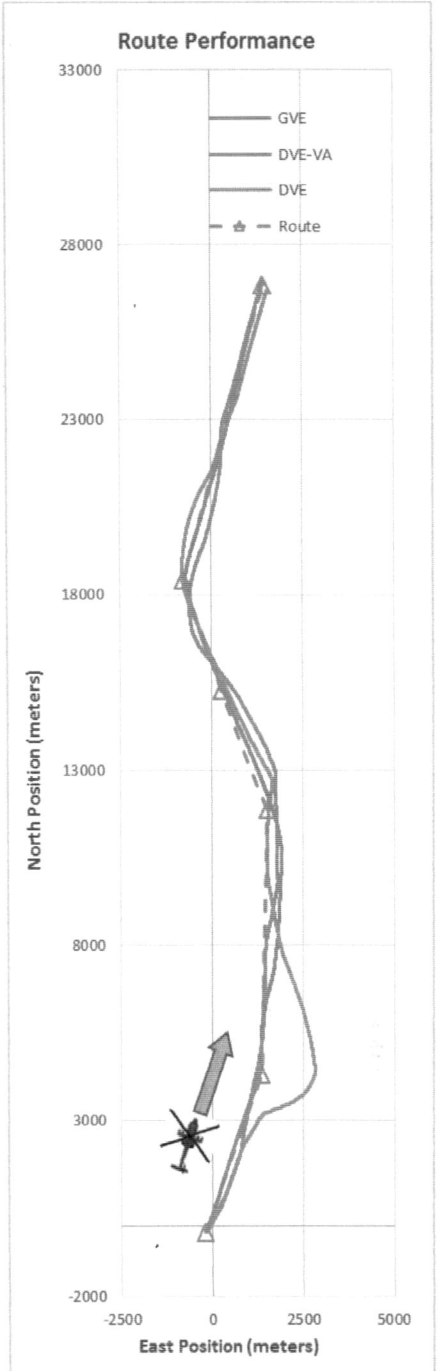

Figure 9. Route Following Enroute to FARP.

Vignette #3 Results

Vignette #3 was included to illustrate the workload associated with landing in confined, unimproved landing areas in and around ground obstacles such as fuel trucks.

Overall, pilot performance and workload/SA ratings for Vignette 3 were positive for GVE flight into the FARP. However, workload ratings ranged from average to high for both DVE and visually aided DVE conditions (with DVE slightly higher) on the go-around phase, whereas the pilots reported that they had poor situational awareness throughout both visually degraded conditions (Figure 12 - Figure 14). The vignette was also intended to highlight the difficulty of keeping power within limits during the go-around, especially when in brownout DVE. However, overtorques did not occur as frequently as expected during the maneuver. Since overtorques are a significant Operations & Sustainment cost driver for the AH-64 and intuition says that they should be expected to worsen in DVE operations, the designed mission task should incorporate maneuvers where the potential for torque exceedances exists. Vignette #4 captures this desire.

Vignette #4 Results

Vignette #4 illustrates the affect DVE has on the ability of the pilot to keep the aircraft within limits and the affect DVE has on performance as well as pilot workload. Figure 10 shows that in GVE, the pilot could easily hold speed and torque constant while enroute to the TIC. However, with visibility degraded, the pilot notably had difficulty keeping speed constant and torque within limits. Even with the HMD 3-D Map visual aiding to compensate, the pilot had overtorques, and 30-40 knot speed excursions were prevalent. These conditions adversely affected his arrival time at the TIC as well as making it difficult for his wingman to fly formation with him (wasting valuable fuel and time). Table 2 shows the route error and torque limit exceedance comparisons between visibility conditions. It was clear that the pilot's ability to monitor and modulate torque within the operating limits was negatively affected by the degraded visibility in DVE. This is seen by the significantly larger number of torque (Q) exceedances (28 vs. 4), and the amount of time in the overtorque region as compared to GVE (187 %-sec vs. 0 %-sec).

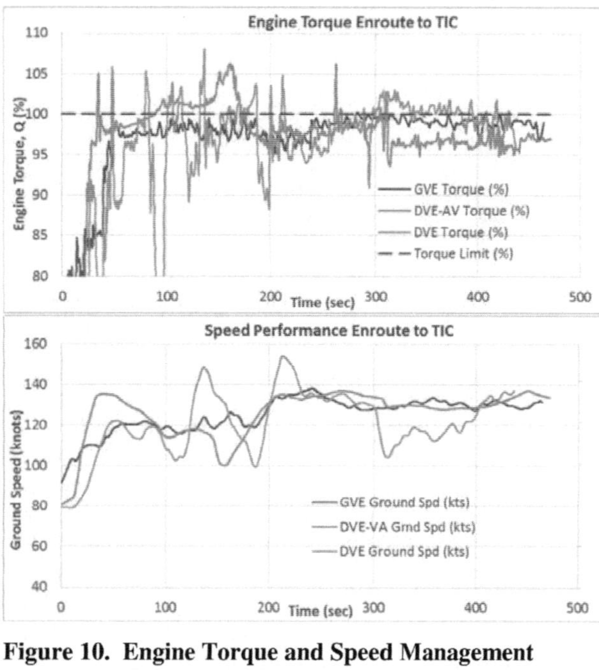

Figure 10. Engine Torque and Speed Management Enroute to TIC.

Condition	Q Exceed Events	Q Exceed Integral (%-s)	Q Exceed Peak (%)	Route err Avg (ft)
GVE	4	0	0	497
DVE(AV)	6	164	6	977
DVE	28	187	8	829

Table 2. Vignette 4.2 Direct to TIC Performance

Pilot provided feedback on this portion of the mission, stating it would greatly reduce the workload and number of overtorques to have a feature where the collective could be placed at maximum torque where he would not have to constantly be monitoring and varying the torque to ensure the limits were not being exceeded. Figure 11 illustrates how the Engine Performance Limit Cueing feature demonstrated on AVMS provides these pilot requested capabilities. For example, the AVMS testing showed how the limit cueing allows the pilot to add power right up to the limit without exceeding it by "riding" the softstop cue; in addition, the capability allows the pilot to get to the limit faster. The pilot can be more aggressive up to the limit because he's confident the limit is not going to be exceeded when he's at or below the softstop. While on the softstop cue, the power remains close to the limit whereas, without the cue, the pilot continually "searches" for the limit position. Without the cue, he sometimes exceeds the limit and sometimes well below it while attention is on other cockpit tasks. This results in lost time and/or performance as shown in the shaded areas of Figure 11. AVMS also demonstrated a reduction in the number of limit exceedances that results in lower maintenance costs[1,2].

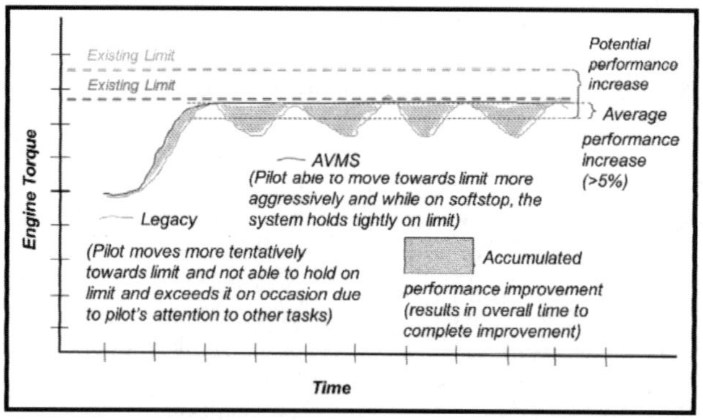

Figure 11. Example showing Existing Limit, Potential Limit, and Ability to Ride the Limit.

Vignette #5 Results

Finally, Vignette #5 was included in the experiment to illustrate the workload associated with maintaining a constant speed and altitude orbit pattern while busy with other mission tasks. Pilot performance, overall, did not deviate across conditions. However, pilots commented that they had to focus more to hold the pattern in the DVE conditions. Thus, pilot performance and workload/situation awareness ratings for Vignette 5 were positive for GVE flight in the Wagon Wheel Orbit, but were poorer for DVE conditions receiving average ratings for both Visually-Aided DVE and DVE conditions (Figure 12-Figure 14).

Results Summary

This section summarizes performance against a number of metrics across all vignettes for each of the three conditions, GVE, DVE and DVE-VA. The metrics examined across all vignettes include situational awareness (both tactical and system), workload and time to complete tasks.

Figure 12 and Figure 13 show the effect of visual conditions on situational awareness. Across all vignettes, tactical situational awareness was highest in the GVE condition, and lowest in the DVE conditions, even when visually aided. However, pilot ratings for Visually-Aided DVE were noticeably improved for enroute vignettes where the pilot was flying to a destination. Pilot feedback provides rationale that the HMD 3-D map visual aiding did not help compensate for the poor visibility in the Lateral Reposition maneuver as much as it did for flying along the route through the terrain. According to the pilots, it was mainly due to the HMD 3-D Map not providing ground surface cueing features to help pick up the relative motion that GVE provides.

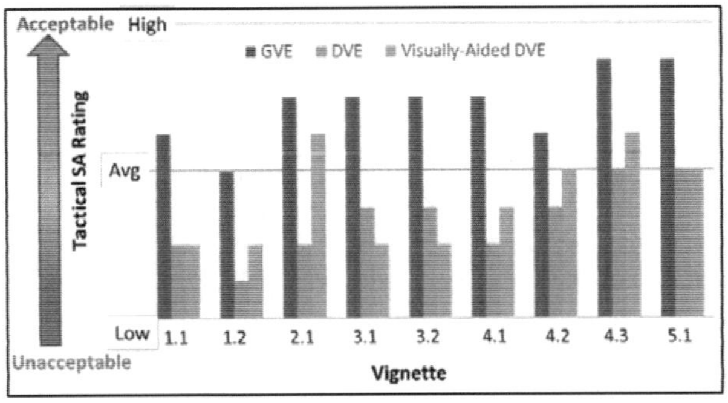

Figure 12. Tactical SA Across All Vignettes/Conditions

Across all vignettes, aircraft systems situational awareness was also highest in the GVE condition, and lowest in the DVE conditions, even when visually aided. Pilot ratings for Visually-Aided DVE were also higher or equal to the DVE for the en-route vignettes where the pilot was flying to a destination (Figure 13).

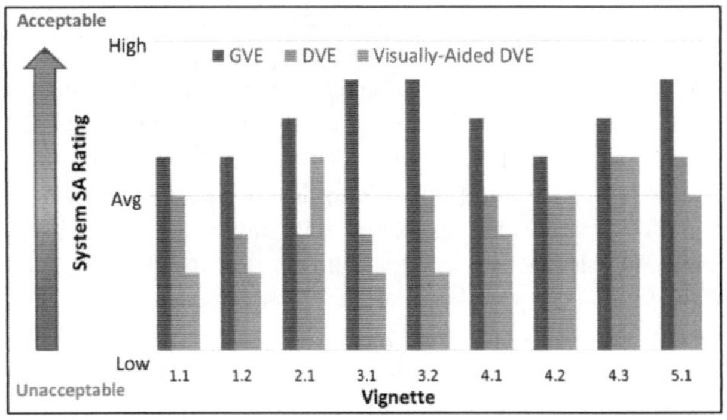

Figure 13. System SA Across All Vignettes/Conditions

Pilot workload ratings showed an increase in workload for both DVE and Visually-Aided DVE (Figure 14). This is consistent with the situational awareness results.

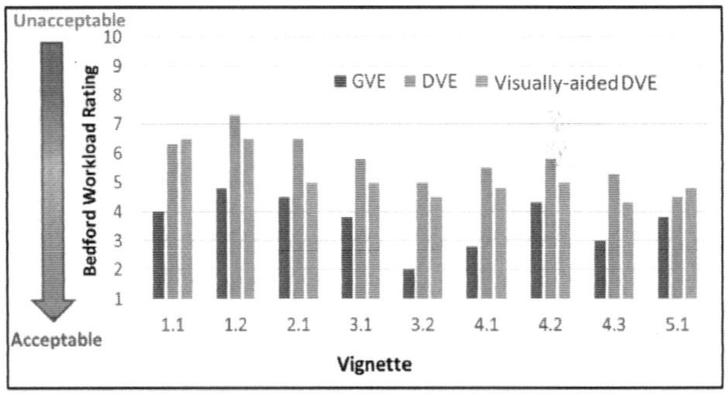

Figure 14. Workload Across All Vignettes/Conditions.

To further show a comparison of workload and situation awareness, the data was normalized into a relative scoring of High, Average, and Low (Figure 15). The TLX was normalized from a 100-point scale down to 3 categories (Objective, Threshold, and Undesired) in order to align with the Bedford (Level 1 or Objective, Level 2 or Threshold, and Level 3 or Undesired).

Bedford/TLX Workload and SA Comparison

Workload Normalization

- Objective (Low) = 1-3 (Bedford/Cooper), and 1-30 (TLX)
- Threshold (Average) = 4-6 (Bedford/Cooper), and 31-60 (TLX)
- Undesired (High) = 7-10 (Bedford/Cooper), and 61-100 (TLX)

SA Ratings were Great, Ok, and Poor

- Objective (High) = Great
- Threshold (Average) = Ok
- Undesired (Low) = Poor

Figure 15. Workload/Situational Awareness Normalization

Figure 16 provides an overall comparison of normalized workload and situation awareness pilot ratings averaged across vignettes, which continues to show the same general trending for situational awareness degradation and increased workload in Visually-Aided DVE and DVE conditions respectively. In addition, when pilots were queried using a 5-point Likert Scale to rate perceived fatigue felt during flight operations under all conditions, the average pilot rating for GVE was 4 (good), DVE was 2.5 (average), and Visually-Aided DVE was 2.2 (average) implying that there is a physical impact on pilots flying in DVE conditions, as well.

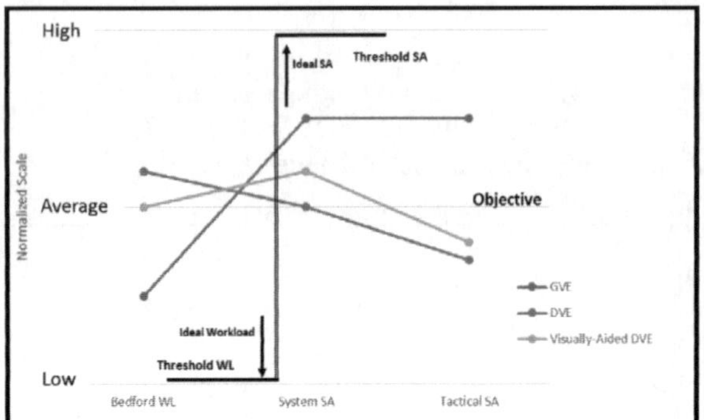

	Bedford	System SA	Tactical SA
GVE	1.5	2.5	2.5
DVE	2.2	2.0	1.7
Visually-Aided DVE	2.0	2.2	1.8

Figure 16. Normalized Comparison of Workload and SA

These overall qualitative ratings support the performance findings across the environmental conditions in this study. In addition, adding a rudimentary visual aiding system did little to improve operating in DVE, thus suggesting that there is a technology gap existing and hence there is significant room for improvement through the application of visual aiding and/or flight control technologies.

Mission execution performance was also measured by timing how long it took for the pilot to perform some of the required tasks. Figure 17 is a Time to Complete (TTC) summary of the lateral maneuvers of Vignette 1 and Figure 18 summarizes the TTC for Vignette #2. These highlight the fact that DVE not only increases the pilot workload but it also results in the pilot taking more time to execute tasks subsequently burning more fuel, arriving late, and having less time on station to perform his attack helicopter mission.

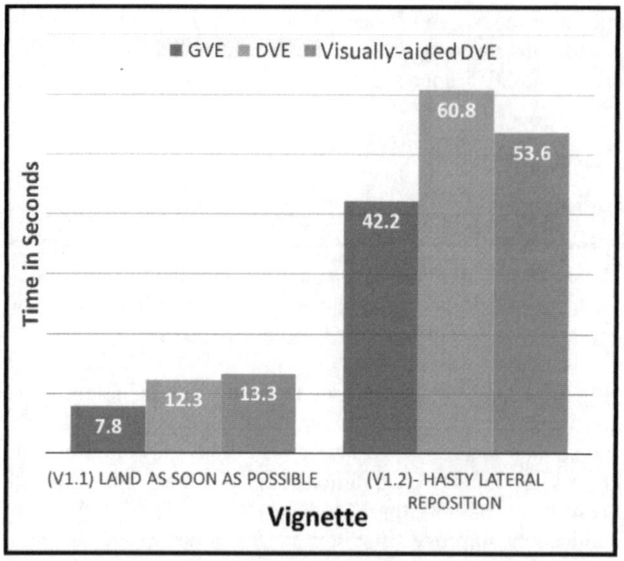

Figure 17. Time to Complete Tasks in Vignette #1

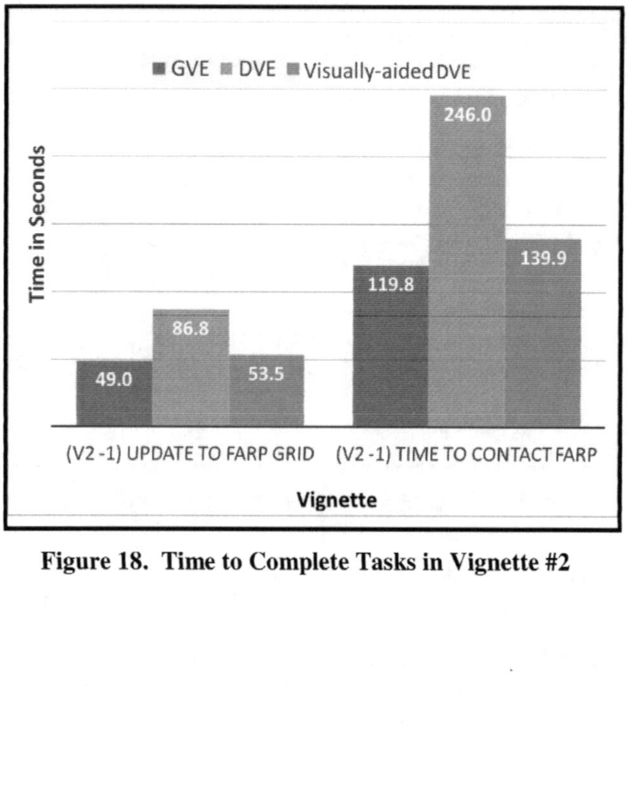

Figure 18. Time to Complete Tasks in Vignette #2

CONCLUDING REMARKS

With a recent emphasis on operating rotorcraft in DVE, this experiment was developed for the AH-64 to measure the effects of operating in DVE on pilot situational awareness, workload, handling qualities and performance while conducting an attack helicopter mission. The results, on average, and across the board, showed that DVE reduced situational awareness, increased work load, reduced handling qualities and worsened performance. In addition, emulating a developmental prototype visual aiding system often made some improvements when operating in DVE.

These results showed the importance of new technologies needed to operate in DVE, especially if there is a desire to approach operating at the same levels of performance as in GVE. Specifically, the results demonstrated the vital need to address all three legs of the desired DVE stool (Figure 19); sensors, displays (cues) and flight controls, ensuring that each leg is sufficiently solid. The paper identified certain technologies expected to be critical to improving the metric deficiencies when operating in DVE, including many tactile cueing and advanced control law technologies developed during the Adaptive Vehicle Management System program. Based on the results of this study, it is projected that the mission test developed in this experiment will serve as a means for evaluating each leg of the stool and what effect they have on the various metrics.

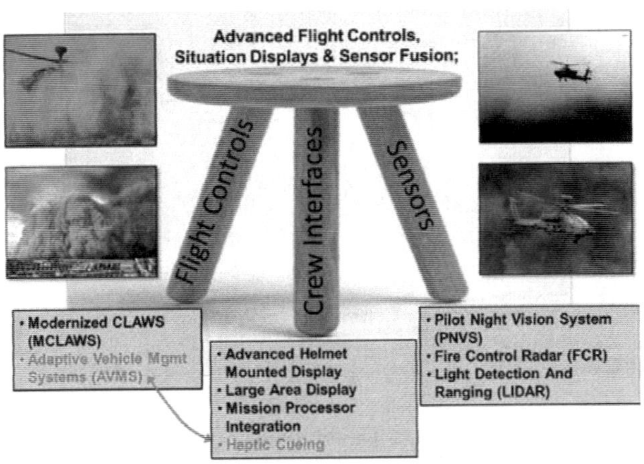

Figure 19. DVE 3-Legged Stool

Author Contact
Gary Klein, gary.d.klein@boeing.com
John Krainski, john.krainski@boeing.com,
Dr. Kendra Befort, kendra.l.befort@boeing.com,
Dr. Russell Enns, russell.enns@boeing.com,

ACKNOWLEDGMENTS

The authors wish to thank Chris Haver for developing the scenarios and test execution, Justin Schaub for simulation technical support, and all of the Boeing test pilots involved.

REFERENCES

[1]Dellert, T., "RDECOM Rotorcraft DVE Mitigation (DVE-M) Program Executive Summary" at the American Helicopter Society 72nd Annual Forum, West Palm Beach, FL, May 17-19, 2016. 72,

[2]Major General Michael Lundy, The Washington Times, "Army not content to 'own the night,' seeks aircraft for all weather conditions", March 31, 2015.

[3]https://www.army.mil/article/108496/Army_aviation_strives_to_own_weather, 15 Aug 2013.

[4]Roscoe, A. and Ellis, G., A Subjective Rating Scale for Assessing Pilot Workload in Flight: A Decade of Practical Use, TR90019, Royal Aeronautical Establishment, 1990.

[5]Likert, Rensis. "A Technique for the Measurement of Attitudes". Archives of Psychology. 140: 1–55, 1932.

[6]Hart, S. and Staveland, L., Development of NASA-TLX (Task Load Index): Results of Empirical and Theoretical Research, Advances in Psychology, Human Mental Workload, Vol. 52, pp. 139-183, Elsevier, 1988.

[7]Anon, ADS-33E-PRF: Aeronautical Design Standard Performance Specification Handling Qualities Requirements for Military Rotorcraft, U.S. Army Aviation and Missile Command, 2000.

[8]Enns, R., Segner, D., Dryfoos, J. "Adaptive Vehicle Management Systems Program Phase II Overview", Proceedings of the American Helicopter Society 72nd Annual Forum, West Palm Beach, FL, May 17-19, 2016.

[9]Klein, G., Kashawlic, B., Enns, R., "Design and Flight Test of an Adaptive Vehicle Management System on an AH-64 Helicopter Demonstrator", Proceedings of the American Helicopter Society 72nd Annual Forum, West Palm Beach, FL, May 17-19, 2016.

Proposed Revisions to the Rotorcraft Failures Section in ADS-33E-PRF

Frank Luria
Aerospace Engineer
Aviation Engineering Directorate
U.S Army Aviation and Missiles and Research, Development, and Engineering Center (AMRDEC)
U.S. Army Research, Development, and Engineering Command
Redstone Arsenal, AL, USA

ABSTRACT

As part of the ADS-33E revision effort, the requirements related to Rotorcraft Failures (Para. 3.1.14) are being revisited. This effort is due, in large part, to the industry standard system safety assessment process that has been adopted for Army rotorcraft development efforts, SAE ARP-4761, "*Guidelines and Methods for Conducting the Safety Assessment Process on Civil, Airborne Systems and Equipment*". Both the rotorcraft Original Equipment Manufacturers (OEMs) and airworthiness authority, US Army Aviation Engineering Directorate (AED), have implemented the ARP-4761 practices into their safety assessment and qualification processes. Proposed changes to the ADS-33E failure requirements that are in line with established safety assessment practice are presented. In addition, additional considerations to improve the hazard identification process crucial to a thorough safety assessment for flight control system (FCS) development are presented.

.

ABBREVIATIONS

ACAH	Attitude Command Attitude Hold
ADS	Aeronautical Design Standard
ADQ	Adequate
AED	Aviation Engineering Directorate
AFCS	Automatic Flight Control System
CLAWS	Control Laws
CON	Controllable
DVE	Degraded Visual Environment
EGI	Embedded Global Positioning System / Inertial Navigation System
FBW	Fly-By-Wire
FCC	Flight Control Computer
FCS	Flight Control System
FCSIL	Flight Control System Integration Laboratory
FHA	Functional Hazard Assessment
FMEA	Failure Modes, Effects Analysis
FTA	Fault Tree Analysis
FVL	Future Vertical Lift
HQ	Handling Qualities
IMC	Instrument Meteorological Conditions
LOC	Loss of Control
MTE	Mission Task Element
OFE	Operational Flight Envelope
OEM	Original Equipment Manufacturers
Pf	Probability of Failure (per flight hour)
SAT	Satisfactory
SOF	Safety of Flight
SSA	System Safety Assessment
STPA	Systems Theoretic Process Analysis
TRC	Translational Rate Command
UCA	Unsafe Control Action

Presented at the AHS Rotorcraft Handling Qualities Technical Meeting, Huntsville, AL, Feb 22-23, 2017.

INTRODUCTION

ADS-33E [Ref. 1] has continued the failure requirements approach initiated by earlier versions of this ADS. In particular, to allow degradations in handling qualities relative to the probabilities of the failure conditions encountered. This approach is sound, and is commonly used in military and commercial system safety assessment. The methodology to verify these requirements are being met can be standardized into a system safety assessment process that is currently best practice in military and commercial aviation. This is the objective of this paper, to walk through this methodology, which is being considered in the revision effort for ADS-33E.

CURRENT REQUIREMENTS

In ADS-33E, Para. 3.1.14 allows a degradation in handling qualities as a result of assessed rotorcraft failure states. This is expanded using two methods. In the first method (Para. 3.1.14.1), the allowable degradation is a function of the probability of encountering the failure state. As an example, failure states that may lead to loss of control must have a very low probability of occurrence, which is expressed in terms of probability of failure per flight hour (Pf). In ADS-33E, these probabilities are specified as shown in Table 1. This method has been derived from the fixed wing specifications, MIL-F-8785C [Ref. 2] and MIL-F-83300 [Ref. 3], where degradations in flying qualities were allowed when the probability of encountering the associated failure states was kept within reasonable bounds. This approach was deemed well-suited to a rigorous failure analysis to determine potential problem areas in flying qualities and was, therefore adopted for ADS-33 [Ref. 4].

One of the issues with this approach is the way Table 1 is applied. According to the ADS-33 Test Guide Para. 3.1.14.1 [Ref. 5], the following requirement, for example, can be inferred from Table 1: Any failure state with a probability greater than 2.5×10^{-3} Pf must be demonstrated to result in Level 1 handling qualities. In addition, any failure state with a probability $2.5 \times 10^{-3} > P_f \geq 2.5 \times 10^{-5}$ per flight hr must be demonstrated to result in Level 2 or better handling qualities within the OFE. In the context of qualification, what does "must be demonstrated to result in Level 1 handling qualities" or "must be demonstrated to result in Level 2 or better handling qualities" mean? It is left for the developer to decide. For example, does "must be demonstrated to result in Level 1 handling qualities" mean that, following the failure, all of the ADS-33 quantitative and qualitative criteria must be met? The Reference 5 Test Guide seems to interpret this as, subsequent to the failure, the aircraft must be in a degraded mode that has been demonstrated to attain that specific level of handling qualities. Therefore, if the probability of the failure is within $2.5 \times 10^{-3} > P_f \geq 2.5 \times 10^{-5}$ per flight hr, the degraded mode must have been demonstrated to meet Level 2 or better. Even with this interpretation, it is not clear what "demonstrated to meet Level 2 or better" means in terms of compliance to the ADS-33E quantitative and qualitative criteria. In addition, the shortcoming of this interpretation is that many failure states do not automatically cause the aircraft to go to a formal degraded mode. Therefore, it will not be known what level the handling qualities are with the failure state occurring while in the full up control mode with the specified failures, rather than a degraded mode for which handling qualities have been demonstrated in some fashion. An alternative approach is discussed in the recommendations section below.

Table 1. Allowable Levels of Handling Qualities Following a System Failure (Table II ADS-33E)

Handling Qualities Level Allowed	Probability of Encountering Within Operational Flight Envelope	Probability of Encountering Within Service Flight Envelope
Level 2 after failure	< 2.5 x 10-3 per flight hr	
Level 3 after failure	< 2.5 x 10-5 per flight hr	< 2.5 x 10-3 per flight hr
Loss of control	< 2.5 x 10-7 per flight hr	

The second method in ADS-33E Para. 3.1.14.2 "assumes that certain failures or combinations of failures will occur regardless of their probability of failure". In these cases, the specific failures will be identified by the Government or qualification authority, along with their respective allowable handling qualities degradation. This may be documented in a system performance specification or other similar requirements document, or in airworthiness qualification requirements.

Additionally, transients following failures are covered in ADS-33E Para. 3.1.14.4. Table 2 shows the allowable rotorcraft transient response subsequent to a failure related to the level of handling qualities. The Test Guide [Ref. 5] seems to indicate that this table should not be used as tailored specification requirements because the criteria in it has not been verified by data. This shortcoming also will be addressed in the recommendations section.

Table 2. Transients Following Failures from ADS-33E-PRF Table III

Level	Flight Condition		
	Hover and Low Speed	Forward Flight	
		Near Earth	Up-and-Away
1	3 deg roll, pitch, yaw 0.05g n_x n_y n_z. No recovery action for 3.0 sec	Both hover & low speed & forward flight up & away requirements apply	Stay within the OFE. No recovery action for 10 sec
2	10 deg attitude change or 0.2g acceleration. No recovery action for 3.0 sec	Both hover & low speed & forward flight up & away requirements apply	Stay within OFE. No recovery action for 5.0 sec
3	24 deg attitude change or 0.4g acceleration. No recovery action for 3.0 sec	Both hover & low speed & forward flight up & away requirements apply	Stay within OFE. No recovery action for 3.0 sec

RECOMMENDATIONS

ADS-33E Para 3.1.14.1 "Allowable Levels Based on Probability" steps a – e did not follow an established commercial or industry standard process to determine whether failures categorized by their effects on levels of handling qualities would meet appropriate probability of failure targets. SAE ARP-4761 "*Guidelines and Methods for Conducting the Safety Assessment Process on Civil, Airborne Systems and Equipment*", Ref. 6, is an established system safety assessment practice that can be used to fulfill the Para. 3.1.14.1 criteria. Since ARP-4761 has been widely adopted in the aviation industry, and is referenced in FAA AC 29.1309 [Ref. 7] as an acceptable means of compliance for the FAA's Equipment, Systems, and Installations systems safety requirements, it is well-suited for the ADS-33 revision effort. OEMs typically use this process for both civilian and military certification efforts for their aircraft.

It is recommended that the ARP-4761 process be used to fulfill the "Allowable Levels Based on Probability" Paragraph 3.1.14.1 in ADS-33E. Steps a-d should be reordered and modified to the following:

Step a. Tabulate all rotorcraft Failure Conditions and their associated failure effects.

Step b. Calculate the probability of encountering each identified Rotorcraft Failure State per flight hour.

Step c. Compute the total probabilities of encountering each of the severity levels. (This total is the sum of the rate of each failure only if the failures are statistically independent)

Step d. Determine whether the Allowable Levels Based on Probability requirements levied on the program have been satisfied.

Discussion of Recommended Steps for ADS-33E Para. 3.1.14.1 Allowable Levels Based on Probability

Step a "Tabulate all rotorcraft Failure Conditions and their associated failure effects" outlines the use of the Functional Hazard Assessment (FHA) in accordance with ARP-4761 for generating the failure conditions and associated effects. This step is really a combination of Para 3.1.14.1 steps a-c in ADS-33E which, due to the FHA process, can be combined into one step. The difference is that the degradation in handling qualities is determined in this step as an outcome of the process of assessing functional hazards and their associated failure effects and severities.

To accomplish this, perform a Functional Hazard Assessment (FHA) in accordance with ARP-4761. The objective of the FHA is to consider aircraft and flight control system functions at the most appropriate levels, and to identify failure conditions and their associated severity classifications, while considering both loss of functions and malfunctions. The severity classifications used in the FHA shall be agreed upon between the design authority and the certification or qualification agency if applicable. These severity classifications shall be based on failure condition effects which include, but are not limited to, handling qualities, pilot workload, and impacts on safety margins and functional capabilities. Typically, the definitions used in Table 3 from ARP-4761 are used in the FHA.

This table is referenced in both ARP-4761 [Ref. 6] and FAA AC 29.1309 [Ref. 7]. The definitions for Severe Major, Major, and Minor severity include the degree of safety margin reduction and impact on crew workload. However, these definitions are somewhat ambiguous and lend themselves to improvement. For example, how is a "large reduction in safety margin" quantified relative to a "significant reduction in safety margin"? In addition, what are the effects of the failures on handling qualities?

As an alternative to the severity definitions in Table 3, a rating scale more focused on effects on handing qualities would be useful. From Appendix 5 of AC 25-7C *FAA Handling Qualities Rating Method* [Ref. 8], the rating scale from Table 4 is used for a pilot task-oriented approach for evaluating handling qualities.

For implementation into the "Allowable Levels Based on Probability" section of ADS-33, these definitions can be used as part of the FHA assessment. For example, to assess the severity level of a specific failure condition, the effects would either be Uncontrollable with Loss of Control (LOC) preventing continued safe flight and landing, Controllable (CON), Adequate (ADQ), or Satisfactory (SAT) per Table 4. To make this assessment easier, Table 5 modifies the AC25-7C rating scale and definitions to also include an allowable probability level based on current guidance from AC 29.1309 and ARP-4761. Therefore, for hazard assessment involving effects on handling qualities, the scale in Table 5 can be used.

Table 3. Failure Condition Severity Definitions taken from ARP-4761 Table 1 [Ref. 6]

Probability (Quantitative)		**1.0**	**1.0E-03**	**Per Flight Hour 1.0E-05**	**1.0E-07**	**1.0E-09**
Probability (Descriptive)	FAA	Probable		Improbable		Extremely Improbable
	JAA	Frequent	Reasonably Probable	Remote	Extremely Remote	Extremely Improbable
Failure Condition Severity Classification	FAA	Minor		Major	Severe Major	Catastrophic
	JAA	Minor		Major	Hazardous	Catastrophic
Failure Condition Effect	FAA & JAA	-slight reduction in safety margins -slight increase in crew workload -some inconvenience to occupants		-significant reduction in safety margins or functional capabilities -significant increase in crew workload or in conditions impairing crew efficiency -some discomfort to occupants	-large reduction in safety margins or functional capabilities -higher workload or physical distress such that the crew could not be relied upon to perform tasks accurately or completely -adverse effects upon occupants	All failure conditions which prevent continued safe flight and landing
Development Assurance Level	ARP 4754	Level D		Level C	Level B	Level A

Note: A "No Safety Effect" Development Assurance Level E exists which may span any probability range

Table 4. Pilot Ratings for Handling Qualities Tests [Ref. 8]

HQR Rating	**Definitions**	**Comparison** **Cooper-Harper**	**ADS-33E Level**	**QUAL**
Satisfactory (SAT)	Full performance criteria met with routine pilot effort and attention	1-3	1	SAT
Adequate (ADQ)	Adequate for continued safe flight and landing; full or specified reduced performance met, but with heightened pilot effort and attention	4-6	2	ACCEPT
Controllable (CON)	Inadequate for continued safe flight and landing, but controllable for return to safe flight condition, as safe flight envelope, and / or reconfiguration so that HQ are at least ADEQUATE	7-8	3	CON

Table 5. Handling Qualities Rating Scale for Hazard Assessment

FHA Severity Level	HANDLING QUALITIES DEGRADATION	HQ Rating	COOPER-HARPER PILOT RATING (CHPR)	ADS33 HQ LEVEL	Allowable Probability Level
CATASTROPHIC	Loss of Control or Dangerous or Intolerable (intense pilot compensation is required for control, or worse)	Uncontrollable with Loss of Control (LOC)	>8.5	**	$< 1.0 \times 10^{-9}$ per flight hr
SEVERE MAJOR	Inadequate for continued safe flight and landing, but controllable for return to a safe flight condition, a safe flight envelope, and/or allows a reconfiguration that provides handling qualities that are at least ADEQUATE.	Controllable (CON)	6.5 to 8.5	3	$< 1.0 \times 10^{-7}$ per flight hr
MAJOR	Adequate for continued safe flight and landing. Full or specified reduced performance met, but with heightened pilot effort and attention.	Adequate (ADQ)	3.5 to 6.5	2	$< 1.0 \times 10^{-5}$ per flight hr
NO EFFECT	Full performance criteria met with routine pilot effort and attention.	Satisfactory (SAT)	<3.5	1	N/A

**Dangerous or Intolerable in accordance with ADS-33 Test Guide [Ref. 5]

The process for generating a FHA is described in Appendix A of ARP-4761, including the identification of failure conditions. Para. A.3.2 in ARP-4761 states, "To generate these failure conditions and assumptions, the analyst needs to understand the possible failure modes." It is crucial for a thorough FHA to rely heavily on flight control and handling qualities subject matter experts. The consequences of not doing this will reduce the likelihood of capturing failure conditions to the extent that an effective system safety assessment would, even leading to a deficiency in generating design assurance levels and safety requirements. This would affect system development and ultimately, the fielding of a safe aircraft.

The following steps are followed to generate a FHA:

- Determine aircraft or system functions
- Determine the failure conditions associated with these functions
- Determine the effects of these failure conditions
- Classify the failure condition effects

Step b "Calculate the probability of encountering each identified Rotorcraft Failure State per flight hour" outlines the use of the Fault Tree Analysis (FTA) in accordance with ARP-4761 which is another industry standard safety artifact that the OEMs are very familiar with, and routinely use in their system safety assessment process. Thus, it is the ideal vehicle to determine the probabilities of the failure conditions identified in the FHA as part of Step a.

The FTA begins with the failure conditions from the FHA, and from a top down perspective, develops these conditions into lower level contributions, such as the examples from ARP-4761 in Figures 1 and 2 [Figures D3 and D4 of Ref. 6]. The probabilities of these lower level contributions are found by various means, perhaps through a Failure Modes and Effects Analysis (FMEA). It is through combining the individual contributions that the probabilities of the failure conditions at the top level are determined.

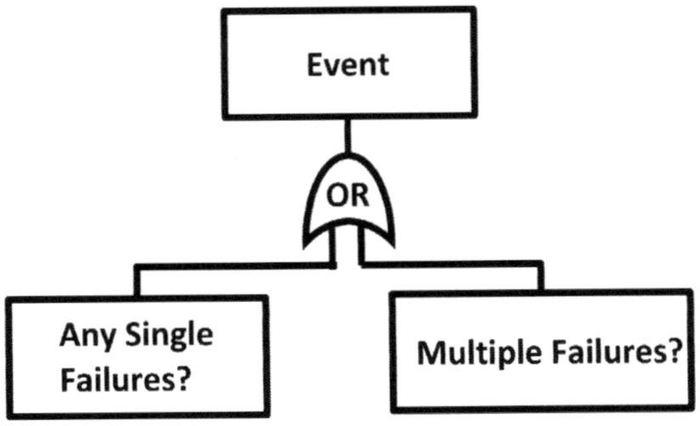

Figure 1. Upper Tier of Fault Tree with Single Threads in the System

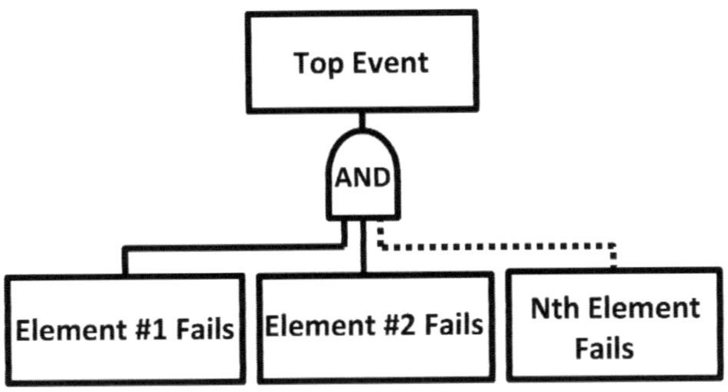

Figure 2. Upper Tier of Fault Tree without Single Threads in the System

The Reference 4 Background Information and User's Guide for ADS-33 offers valid guidance to assessing the failure conditions relative to handling qualities. Rather than determining failure states simply as the summation of component failures, two types of effects are discussed. They are that one failure in a certain mode may itself induce other failures in the system, and also that one event may cause loss of more than one part of the system. The first effect is relevant because of the many control modes possible with advanced digital flight control systems. All of the failure conditions for each control mode would need to be assessed in the various operating conditions that are anticipated. The second effect is usually covered by the FHA and FTA process, in that the failure conditions from a single event are accounted for by all of the functions that are affected. Therefore, these failure conditions will be traceable to all of the functions accounted for in the FHA, FTA, and subsequent safety analyses.

Step c "Compute the total probabilities of encountering each of the severity levels" involves grouping the failure conditions found in Step a. by their severities, and summing their associated probabilities from Step b. The result of this are the total probabilities of encountering each of the categories of handling qualities degradations in Table 5 as they relate to the severities used for the FHA in Step a.

Step d "Determine whether the Allowable Levels Based on Probability requirements levied on the program have been satisfied" involves determining whether the probability of failure requirements have been achieved. If not, an explanation of the shortcomings needs to be provided, including an assessment of the residual risks of exposure to the higher than required probability of failures. The specific documentation for this assessment is dependent on the qualification requirements, but can be contained in the System Safety Assessment (SSA) in accordance with ARP-4761 as an example.

Page 16 of the Reference 5 Test Guide for ADS-33E states that "Flight testing of each failed state is required to obtain the data to determine criterion compliance (Level 1, 2, or 3)." This would pertain to all of the combinations of failures that have a probability of occurrence of greater than 2.5×10^{-7} per flight hour. This is, obviously, a difficult task if only for the high number of potential test points and the mechanism to induce these failure states. In addition, the effects of these failures could have high hazard severity levels, which would make testing specific failure states of high risk. The development of modern flight control systems and advanced control laws involves use of high fidelity flight control system integration labs (FCSIL). This environment is ideal for performing engineering testing to simulate and determine the transient and steady state effects of these failure states on an aircraft. The author recommends use of the FCSIL testing in place of flight testing, particularly with the failure states of high consequence. This does not negate the need to do certain failure modes or degraded modes testing on an aircraft, but rather keeps the overall verification of safety requirements acceptable in terms of risk management, and also efficiency in the flight test program.

Discussion of Recommendations for ADS-33E Para. 3.1.14.2 "Failures or Combinations Occurring Regardless of Probability"

This paragraph in ADS-33E states, "The allowable Level of flying qualities for each Specific Failure will be specified by the Government." As was discussed previously, simply stating the level of flying qualities is ambiguous and should be further defined in the appropriate specification or qualification documents. A better approach would be to utilize the Table 5 criteria for the allowable levels of flying qualities degradation, and specify which levels are acceptable in the specification or qualification documentation.

Discussion of Recommendations for ADS-33E Para. 3.1.14.4 "Transients following Failures"

The requirements for transients following failures should be covered within the ADS-33E Para. 3.1.14.1 Allowable Levels Based on Probability. The failures using the ARP-4761 process are assessed from both a transient and steady-state effect on the rotorcraft. However, there may be cases where the transient response to the failure state needs to be evaluated during testing, either in a FCSIL or a flight test environment.

It is the author's recommendation that Table III in ADS-33E be replaced with a pilot rating scale which evaluates the effect of the failure and the pilot's ability to recover from the failure. The current methodology of evaluating transients following failures using Table III in ADS-33E is, in the words of the Test Guide [Ref. 5], "not supported by data, and no data is currently available to update the table."

Therefore, in the absence of this data, or perhaps sufficient data to support a quantitative evaluation scale, a qualitative rating scale would be more appropriate to evaluate transients following failures. Table 5 can be used for this purpose, which is also being proposed to assess the failure conditions and their effects on handling qualities.

If there is a desire to evaluate these transients with more granularity, the Failure/Recovery rating scale from Hindson et al, Reference 9, can be used. This scale, Fig. 3, has been used in several test programs [References 10 and 11] successfully to evaluate the effect of the failure and the pilot's ability to recover the aircraft.

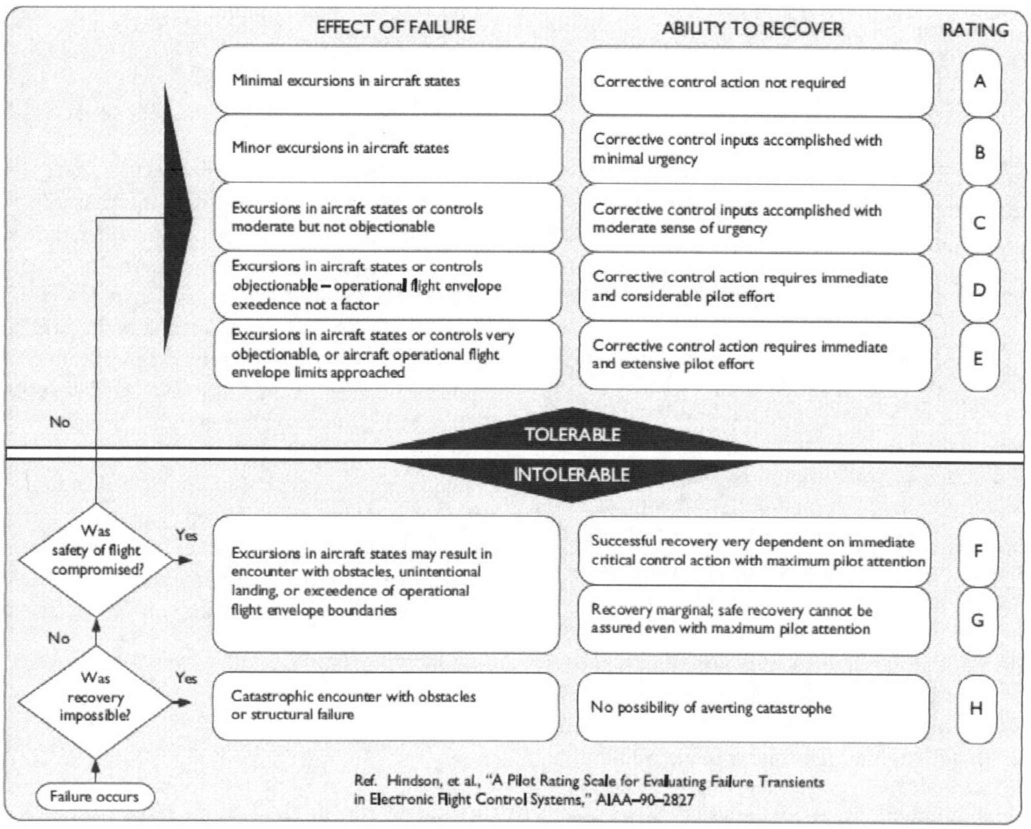

Figure 3. Failure / Recover Rating Scale [Ref. 9]

ADDITONAL CONSIDERATIONS TO FAILURE REQUIREMENTS INVOLVING HANDLING QUALITIES

1. How to identify additional hazards besides from an FHA with its emphasis on failures?

An event, not necessarily a failure event but one that has unforeseen consequences, may have catastrophic consequences. This is one of the strengths of the Systems Theoretic Process Analysis (STPA) methodology of safety analysis. In her book "*Engineering a Safer World*" [Ref. 12], Dr. Nancy Leveson explains the emphasis of designing an effective control structure that eliminates or reduces adverse events, rather than simply preventing failures.

Including unsafe and unexpected interactions among components, or assessing latent failures and operating in conditions other than those that the system was designed to operate in would add thoroughness and robustness to the standard practice of developing failure conditions using the FHA process. STPA treats safety as a control problem, not a component failure problem, and not only identifies and addresses component failures and their effects on the system, but also design flaws and other unsafe actions [Ref. 13]. STPA does not involve any of the standard practice of ARP-4761; however, it can be effective in augmenting an existing hazard assessment process.

The portion of STPA that can augment the FHA process involves the assessment of unsafe control actions, namely to identify the potential for inadequate control of the system that could lead to a hazardous state. Hazardous states result from inadequate control or enforcement of the safety constraints [Ref. 12], which can occur because:

a) A control action required for safety is *not* provided or *not* followed.
b) An unsafe control action *is* provided.
c) A potentially safe control action is provided too early or too late, that is, at the wrong time or in the wrong sequence.
d) A control action required for safety is stopped too soon or applied too long.

An example of this examination of unsafe control actions was performed in Reference 14. The control action involved the automatic flight control system (AFCS) commanding the UH60 Black Hawk variable incidence stabilator installed on the tail rotor pylon. The stabilator provides pitch stability in forward flight. Table 6 contains an extract of this work. The first column lists control actions that can be given by the controller, and the four following columns list how those control actions could be hazardous in four general categories. These hazardous control actions are referred to as unsafe control actions (UCA).

Table 6. STPA Step 1, Assessment of Unsafe Control Actions for UH60 Black Hawk Stabilator System

Control Action	Not Providing Causes Hazard	Providing Causes Hazard	Wrong Timing or Order Causes Hazard	Stop Too Soon or Applied Too Long
	UNSAFE CONTROL ACTIONS			
Stabilator command given (slew up, slew down)	Unsafe control moment as aircraft accels	Incorrect stab postn for given airspeed	Oscillatory mvmt if control loop unstable (small) NOT A HAZARD	Increase workld if slow to command down for quick decel NOT A HAZARD
	Unexpected pitch trim attitude during decel NOT A HAZARD	Unsafe control moment for given flight condition	Differing commands to each actuator results in incorr stab position for given airspeed	Incorrect stab postn for given airspeed

The next step in the STPA process is to determine how each of the potentially unsafe control actions identified in Table 6 could occur. An analysis of these causes would be used to determine if these scenarios are mitigated or eliminated in the proposed design. If they are not, then a design change is warranted.

The STPA process takes these findings and develops safety constraints for design of the system. One of the goals of the STPA is to ensure the safety constraints are enforced in the system design, manufacturing, and operations. As can be seen from Table 6, the steps taken for the STPA can add to the FHA being performed by considering contributing factors other than failures, to include such things as unsafe or unexpected interactions among system components or functions. In the case above, it can consider functions that were not originally designed for in the development of the flight control system.

2. Can standard criteria be added to the hazard identification process?

Consideration should be given to certain flight control system criteria to be included as part of the system safety assessment process. An example would be the stability margin performance in terms of gain and phase margins. AS94900 [Table 2 of Ref. 15], contains the criteria shown in Table 7. Meeting these criteria is not only sound design practice, but helps ensure stability of the rotorcraft throughout the potential myriad of control modes available.

Table 7. Gain Margin (GM) and Phase Margin (PM) Requirements from AS94900 [Ref. 15]

Mode Frequency (Hz)	Airspeed	
	Below V_{OMIN}	V_{OMIN} To V_{OMAX}
$f_M < 0.06$	GM = 6dB (No Phase Reqmt. Below V_{OMIN})	GM = ±4.5 dB PM = ±30 degrees
$0.06 \leq f_M$ First Aeroelastic Mode	GM = 6dB (No Phase Reqmt. Below V_{OMIN})	GM = ±6.0 dB PM = ±45 degrees
$f_M >$ First Aeroelastic Mode	GM = 6dB (No Phase Reqmt. Below V_{OMIN})	GM = ±8.0 dB PM = ±60 degrees

The hazards associated with low stability margins are not typically included in a Functional Hazard Assessment. One reason is that they are not the result of a failure condition. Additionally, instabilities in the control system are typically addressed during the control system design and analysis phases, during FCSIL testing or even during flight testing. Any required fixes are made during those phases in a tightly controlled, highly regulated environment, thereby minimizing hazards.

However, including stability margins in the safety assessment would keep them visible throughout the development and verification process. This would also benefit the qualification authority, especially if the stability margin criteria were included in the performance specification.

Future Vertical Lift (FVL), advanced technology rotorcraft will invariably incorporate fly-by-wire complex flight controls with flight control or vehicle management computers, or even distributed units throughout the aircraft performing command processing. Various hierarchy of control modes are envisioned, for various purposes. The stability margins of the various control loops being closed, should be tracked, with marginal performance being flagged and potentially addressed early on in a program.

As an example, Figure 4 shows results from CONDUIT® for the UH-60M Black Hawk. Both the pitch axis gain and phase margins for the standard response type, rate command with flight path stabilization, and also with hover position hold engaged are shown. A flight control system with higher augmentation response types required by ADS-33, such as Attitude Command Attitude Hold (ACAH) or Translational Rate Command (TRC), should have the stability margin criteria included as part of tracking potential hazards associated with normal operation in adverse operating environments.

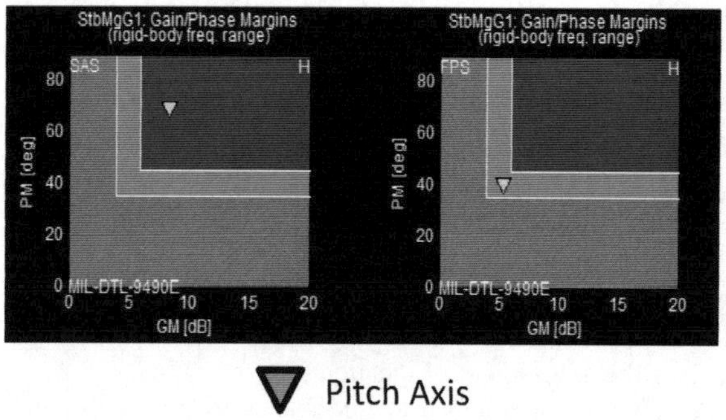

Figure 4. Stability Margin Estimates from CONDUIT® Spec Evaluation for UH-60M Model at Hover

3. Functionally Unimaginable Hazards

As was mentioned earlier, the standard FHA approach is to consider aircraft and flight control system functions at the most appropriate levels, and to identify failure conditions and their associated severity classifications, while considering both loss of functions and malfunctions. However, hazard identification may well include hazards that are not the result of system or component failures. There is a category of these hazards that are hard or even impossible to identify using traditional techniques, such hazards are sometimes called Functionally Unimaginable Hazards [Ref. 16]. The difficulty is determining these hazards in a systematic, creative way. The Reference 17 guidance, *"Guidelines for the Identification of Hazards, How to make Unimaginable Hazards Imaginable?"*, outlines such a methodology. It involves the use of hazard identification brainstorming sessions with operational and relevant experts in a controlled environment, and is meant to complement the standard FHA approach to hazard identification and assessment.

The following steps comprise this methodology which starts with the goal of safe operation of the system [Ref. 17]:

1. Preparing a hazard identification brainstorm
    a. Select and arrange the participants, especially the operational experts;
    b. Prepare an oversight presentation of the operation;
    c. Prepare the brainstorming approach;
    d. Prepare the content of the hazard identification (presentation and hazard categorization)
    e. Consider practical aspects of the hazard identification brainstorming.

2. Perform a hazard identification brainstorm
    a. Take strict care that the basic rules of brainstorming are respected (as many hazards as possible with no analysis/ criticism)
    b. Make short notes of the mentioned hazards on the flip over (easel) using the format "hazard identification (number) and short description" and watch that hazards are correctly understood
    c. Take subtle care that "all" aspects of the operation and possible hazard categories are covered
    d. Apply short breaks *before* productivity drops significantly, such that the participants can free their memory.

Note that for "2c", good preparation by the moderator will help cover "all" aspects by: drawing attention to a not yet covered aspect of the operation on the overview sheet; asking the participants whether there could be hazards related to a specific conflict type; and asking the participants to look for hazards related to a specific hazard category [Ref. 17]

3. Follow-up
    a. Within a few days make and distribute the minutes of the meeting with the numbered list of hazards among the participants, asking them for corrections and additions
    b. Check the effectiveness of the brainstorm
    c. Decide if additional hazard identification brainstorming is necessary

It is thought that the standard FHA functional approach as compared to the brainstorming approach to hazard identification will yield different kinds of subsets of hazards associated with the system operation. This can be seen in Figure 5 where the different hazard subsets are depicted in the solution space of all hazards [Ref. 17]. As seen from this figure, a conclusion would be that a more thorough job of hazard identification is made using both approaches because each has its areas of coverage, which do not necessarily overlap.

One last suggestion is that the brainstorming method be performed BEFORE the standard FHA process. This is due to the fact that the brainstorm process, if done after the FHA, will fixate on the already established functional hazards. This may potentially restrict the value of utilizing this complementary method of identifying hazards to carry forward into the overall safety assessment process. More information can be found in the Reference 17 guidance material.

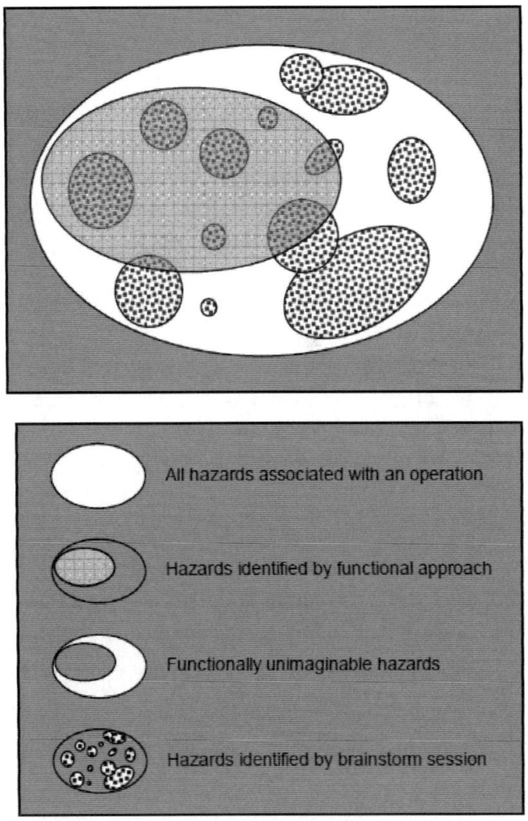

Figure 5. Functional and Brainstorming Approaches Yield Different Hazard Subsets

CONCLUDING RECOMMENDATIONS

The Rotorcraft Failures section in ADS-33E has not been updated in many years. Commercial and military standardized processes in the area of system safety assessment can be applied for compliance to the ADS-33E requirements. Specifically, the SAE ARP-4761 system safety assessment guidance can be tailored comply with ADS-33E. This necessitates some updates to be made to ADS-33, which would follow these general steps:

a. Tabulate all rotorcraft Failure Conditions and their associated failure effects.

b. Calculate the probability of encountering each identified Rotorcraft Failure State per flight hour.

c. Compute the total probabilities of encountering each of the severity levels. (This total is the sum of the rate of each failure only if the failures are statistically independent)

d. Determine whether the Allowable Levels Based on Probability requirements levied on the program have been satisfied.

There are other areas of improvement related to the system safety. While they may not specifically affect ADS-33E, they can be considered during the development phase of a handling qualities improvement program to augment the existing safety process. These three areas are:

1. Use of the System Theoretic Process Analysis (STPA) to do a more thorough job of hazard identification and developing safety requirements for a flight control system upgrade. The emphasis in STPA is to look at safety as a control problem which addresses not only failure conditions, but also unsafe and unanticipated interactions among subsystems leading to a thorough analysis of all hazardous conditions.
2. Use of flight control system design criteria from AS94900 as an indicator of potential deficient design that could lead to hazardous operation of the rotorcraft in difficult operating environments.
3. Use of a methodology to conduct brainstorming sessions to identify additional hazardous conditions in the development of aircraft systems that would normally not be identified through traditional methods. These additional hazards are categorized as Functionally Unimaginable Hazards.

REFERENCES

1. Anon., "Aeronautical Design Standard Performance Specification Handling Qualities Requirements for Military Rotorcraft", ADS-33E-PRF, 21 Mar 2000.

2. Anon., "Military Specification, Flying Qualities of Piloted Airplanes," MIL-F-8785C, 5 November 1980.

3. Anon., "Military Specification, Flying Qualities of Piloted V/STOL Aircraft," MIL-F-83300, December 1970.

4. Hoh, Roger H., David, G. Mitchell, Bimal L. Aponso, David L. Key, Chris L. Blanken, "Background Information and Users Guide for Handling Qualities Requirements for Military Rotorcraft", USAAVSCOM Technical Report 89-A-008, December 1989.

5. Blanken, C.L., Hoh, R.H., Mitchell, D.G., and Key, D.L., "Test Guide for ADS-33E-PRF", Aviation and Missile Research, Development, and Engineering Center Special Report AMR-AF-08-, July 2008.

6. SAE ARP-4761 *Guidelines and Methods for Conducting the Safety Assessment Process on Civil Airborne Systems and Equipment*, Dec 1996.

7. FAA Advisory Circular 29-2C Change 5, Para. 29.1309 *Equipment, Systems, and Installations*, 13 June 2008.

8. FAA Advisory Circular 25-7C, *FAA Handling Qualities Rating Method*, 16 Oct 2012.

9. Hindson, W. S., Eshow, M. M., and Schroeder, J.A., "A Pilot Rating Scale for Evaluating Failure Transients in Electronic Flight Control Systems," AIAA-90-2827, AIAA Atmospheric Flight Mechanics Conference, Portland, OR, August 1990.

10. Weakley, J.M., Kleinhesselink, K.M., Mason, D.H., Mitchell, D.G., "Simulation Evaluation of V-22 Degraded Mode Flying Qualities," American Helicopter Society 59th Annual Forum, Phoenix AZ, May 2003.

11. Kalinowski, K.F., Tucker G.E., Moralez E., "In-Flight Validation of a Pilot Rating Scale for Evaluating Failure Transients in Electronic Flight Control Systems" American Helicopter Society 62nd Annual Forum, Phoenix AZ, May 2006.

12. Leveson, N.G., *Engineering a Safer World: Systems Thinking Applied to Safety*, Massachusetts Institute of Technology (MIT) Press, Cambridge MA, 2011.

13. Albrecht, B., Arterburn, D., Horney, D., Schneider, J., Abel, B., "A New Approach to Hazard Analysis for Rotorcraft", American Helicopter Society Development, Affordability and Qualification of Complex Systems Specialists' Meeting, Huntsville AL, Feb. 2016.

14. Luria, F., "Flight Test Approval for Flight Control System & Handling Qualities Upgrade Development Efforts", American Helicopter Society Rotorcraft Handling Qualities Specialists' Meeting, Huntsville AL, Feb. 2014.

15. SAE Aerospace Standard AS 94900, *Aerospace - Flight Control Systems - Design, Installation and Test of Piloted Military Aircraft, General Specification For,* July 2007.

16. Safety Management System and Safety Culture Working Group (SMS WG), "Guidance on Hazards Identification", ECAST European Strategic Safety Initiative, March 2009

17. De Jong, H.H., "Guidelines for the Identification of Hazards, How to make Unimaginable Hazards Imaginable?", NLR-CR-2004-094, EUROCONTROL European Organisation for the Safety of Air Navigation, March 2004.

AUTHOR INDEX